Deepen Your Mind

作者序

❖ 大數據如何入門？

大數據的時代，各行各業都在試著往大數據前進，將大數據導入企業，那最重要的是什麼呢？模型嗎？筆者認為是資料，沒有資料，一切都是空談，因此基礎工程、基礎建設尤為重要，那要如何完善基礎工程呢？這點就是本書談的核心，資料工程。

❖ 大數據產品

本書從最基本的資料收集、爬蟲開始，到資料庫、RESTful API、分散式，最後走到視覺化，完整的呈現，大數據產品的發展過程。筆者除了做開源資料，為大數據盡一份心力外，也希望將過程，寫成一本書，分享給大眾，希望提供一個入門磚，且不只是入門，本書所用到的技術，已經到 Senior 等級。另外，本書不單純以技術為主軸，而是引入真實案例、真實情境，讓讀者更能體會到，為什麼要使用這些技術，是為了解決什麼樣的問題。

❖ Side Project

工程師、分析師，除了專研技術以外，發展個人的 Side Project，也漸漸成為主流，筆者的專案，FinMind，也是從 Side Project 慢慢發展起來，目前在 GitHub 得到 1,500 stars，算是得到一些認可。開發 Side Project 有什麼好處呢？除了對履歷上加分外，持續開發，接觸的層面會更多更廣，對於架構設計、產品開發上，更加有經驗。且不只是接觸技術，而是有真實的案例，可以證明，你會這項技術，並應用在真實場景。另外，工程師最大的優勢是，有能力做出產品，甚至是分析能力，都有可能幫助讀者，創造出個人產品、品牌，筆者希望能借助本書，讓更多讀者進入這個領域，更多讀者開始建立個人 Project、產品、品牌。

最後，感謝深智數位出版社，邀請筆者撰寫這本書，讓筆者有機會，分享本書內容、產品開發經驗、大數據技術，給各位讀者。

作者簡介

林子軒，Sam，目前任職 17 LIVE 資料工程師。擅長資料工程、資料分析，希望對 Python 社群、大數據領域，提供一份心力。

經歷

- 17 LIVE 資料工程師。
- 曾任職永豐金證券，軟體工程師。
- 曾於 Open UP Summit 2019，擔任 Speaker。
- 曾任職 Tripresso，資料工程師。
- 東華研究所，應用數學碩士。

FinMind

- https://github.com/FinMind/FinMind
- https://finmindtrade.com/

個人 GitHub

- https://github.com/linsamtw

Email

- samlin266118@gmail.com

如對本書有疑問，歡迎寄信到以上信箱。

筆者除了本書內容外，GitHub 上也有寫文章介紹，兩篇 Kaggle 競賽經驗，分別是生產線分析、庫存需求預設，如對以上有興趣，也可寄信到以上信箱。

目錄

06 資料提供—RESTful API 設計

07 容器管理工具 Docker

第 2 篇　產品迭代 -- 測試運維

08　自動化測試

09　CICD 持續性整合、部屬

第 3 篇　API 產品上線

10 API 服務網址

第 4 篇　資料視覺化

11 視覺化工具

A 結論

第 1 篇
資料工程 ETL

Chapter

01

本書介紹

本書主要介紹在 GitHub 上，獲得 1,500 stars 的開源專案 --- FinMind 的架構，獨家解析內部使用的技術。該專案以大數據、資料工程為起點，開發出許多數據服務，包含 Python Package、API、視覺化分析、網站等，呈現產品從 0 到 1 的完整過程，看完此書，你將有能力打造自己的產品、作品，可以作為一個進入大數據領域的入門磚。

讀完本書後，你將完成一個以 Data 為主的 Side Project，包含資料收集、資料庫、API、視覺化呈現，一整套從 0 到 1 的作品，細節如下。

❑ 資料收集

1. Python 環境、套件管理工具 pipenv。
2. 證交所、櫃買中心、期交所爬蟲。
3. 使用 RabbitMQ、Flower、Celery 分散式爬蟲。

❑ 資料庫

4. 使用 Docker 架設 MySQL 資料庫。

❑ Api

5. 使用 FastAPI 撰寫 Restful Api，並用 Docker 包裝成 Image。
6. 使用 No-IP 申請免費網址、使用 Let's Encrypt 免費的 SSL 憑證。

❑ 視覺化

7. Redash 業界視覺化工具。

❑ 其它

8. Linode 雲端。
9. Docker、Docker-Compose 與 Docker Swarm 現代化集群架構。
10. 爬蟲、API Unit Test 單元測試。
11. 使用 GitLab-CI、Gitlab-Runner 建立 CICD 自動化流程。

對於初學者來說，擁有以上技能，對於尋找初階 Data 相關工作，是很容易的（讀者可以搜尋 data 相關職缺）。而對於已經是 Junior，並想往 Senior 發展的讀者，Docker、Docker Swarm、Gitlab-CI、CICD、Unit

Test 等工具,對讀者來說非常有幫助,可以協助完成高品質的程式,並藉由持續開發 Side Project,累積相關經驗。

筆者初期是資料科學背景,深深體會沒有資料的問題,因此慢慢走向資料工程,自動化收集資料,中期希望將分析結果呈現給更多人觀看,單純視覺化不夠,又跑去寫 Web,最後為了達到產品穩定性,做了許多維運的工,書中將會介紹每一步的過程。

FinMind 介紹

近年大數據技術蓬勃發展,許多人都嚮往成為資料科學家,而在資料科學的道路上,如果沒有好的資料工程作為基礎,沒有好的資料源,就像一台很強的火箭,卻沒有燃料,也是空談。

因此,FinMind 收集各種股市資料,並提供一般大眾直接拿取資料,可以寫程式下載資料,也可直接在網站上下載,解決了火箭燃料的問題,而本書中,會解析 FinMind 使用的資料工程技術。

該開源專案主要是使用分散式架構、爬蟲,進行收集資料、資料清洗、存入資料庫,並開發 API 供一般大眾使用。書中除了介紹最基本的爬蟲外,分散式、資料庫架設、API 開發、Loading Balance 等,都會一併介紹,並使用 Docker 進行多台機器的控管,在寫書的當下,FinMind 使用 9 台雲端機器,而如何控管這些機器,也會是書中介紹的重點之一。

產品架構圖

下圖是 FinMind 產品開發中,所使用到的技術,從最初的資料收集,到最後的 API、資料分析、視覺化、網站呈現、金流串接。

註 1：一般程式開發工具，常見的都是以黑底為主，而由於書本印刷原因，將以白底呈現，這點請讀者特別留意。

註 2：一般程式的排版，長度會以 80 為主，本書因為排版問題，長度會以 40 為主。

開發環境

2.1　開發環境重要性

首先，先介紹開發環境設定，有適合的作業系統、適合的開發工具，會讓你事半功倍，以下將介紹業界常用的開發工具。

2.2　Linux 作業系統

建議開發環境使用 Linux，市面上大多數服務、網站，都使用 Linux 系統，且雲端大多都提供 Linux 環境，因此直接在 Linux 下做開發，這樣對於產品部屬上，不容易遇到環境問題。如果只有 Windows 環境的話，下一個章節，會介紹如何在 Windows 環境下開發。

本書建議 Linux 的讀者使用 Ubuntu 18.04，這算是最知名的 Linux 作業系統之一，且相對來說，發生問題時，網路上的資源相對較多。

2.3 Windows 作業系統

過去如果要在 Windows 上使用 Linux 環境開發的話，大多會借助 VM 等虛擬機器，但多數虛擬機效能都不好。近年微軟開發出 WSL（Windows Subsystem for Linux）這套工具，可以讓開發者，在 Windows 上直接使用 Linux 環境，以下將介紹如何在 Windows 使用 WSL。

1. 準備 win10 專業版。

2. 開啟 [控制台]，點選 [程式和功能]

3. [開啟或關閉 Windows 功能]

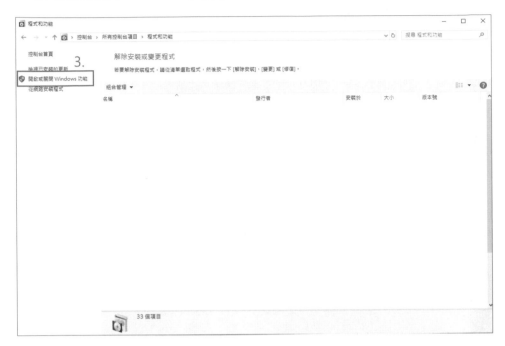

4. 勾選,適用於 Linux 的 Windows 子系統

以上，已經啟用 Windows 中 Linux 子系統功能，接下來將安裝其中一個 Linux 系統 Ubuntu。

5. 進入 Microsoft Store

6. 搜尋 Ubuntu

7. 安裝 Ubuntu

8. 安裝完後，點啟動

9. 啟動完後，就會顯示此畫面，就有 Linux 環境啦！

2.4 Mac 作業系統

如果是 Mac 的讀者，由於 Mac 操作上與 Linux 相近，因此不需要做額外設定，可直接進入 2.5 章節。

2.5 Python 開發工具 VS Code

好的 IDE 在開發上會事半功倍，本書使用 VS Code 這套 IDE 開發工具，因為它功能豐富，請讀者到以下連結下載 VS Code，本書使用的版本是 1.41.1。

https://code.visualstudio.com/updates/v1_41

Windows 的讀者，請選擇 System，Mac 直接選擇 64 bit，Ubuntu 讀者選擇 deb 後安裝。

安裝完成後，如果是 Ubuntu、Mac 的讀者，就可以直接在 VS Code 的
TERMINAL 使用原生 Mac、Ubuntu 指令，如果是 Windows 環境，VS
Code 需要做以下設定，在右下角，點選 **Select Default Shell**，並選擇
bash，之後點選垃圾桶，重新打開 TERMINAL，就會是 bash 了。

VS Code 全名 Visual Studio Code，是由微軟開發的 IDE 編輯器，一般
Python 初學者，可能會選用 Spyder、Jupyter、Sublime，進階一點使用
PyCharm 、VS Code，本書將會以 VS Code 為主做開發，以下將介紹 VS
Code 幾個優點。

❑ 輕量

相對於 PyCharm 來說，VS Code 相對輕量，開啟速度快，可以加快進入
開發環境中，本書會開發多個專案，需要開啟多個 IDE，因此效能也是考
量之一。

❏ 插件

VS Code 有豐富的插件，可額外下載一些工具，對你的 VS Code 進行擴充，例如你要跑測試、觀看 Git Graph、甚至使用 Tabnine 這項 AI Autocomplete 自動補齊的工具。

❑ 測試

在開發過程中,一定會經歷需要跑測試的環節,初學者可能不清楚什麼是測試,這在接下來的章節中會提到。而 Spyder、Jupyter、Sublime 都沒有測試的功能。

3.1　為什麼先介紹 Docker ？

在開發 Python 專案時，初期可能只會停留在寫 code 階段，到了中期時，你會需要許多工具來幫助你。

以資料工程為例，初期收集資料，需架設資料庫來做儲存，如 MySQL，進階一點做分散式爬蟲，如 RabbitMQ、Flower，而安裝這些工具，需要一系列繁瑣的過程，如果你在 Linux，那安裝過程稍微好一點，如果在 Windows，更加繁瑣，借助 Docker，可以一行指令架好這些工具，在 4.3 將會介紹如何一行指令安裝資料庫。

3.2　什麼是 Docker？

在使用 Docker 之前，本書先介紹什麼是 Docker。

Docker 類似於虛擬機的概念，幫你分離出獨立環境，在產品開發上，可以在本地端模擬實際部屬的情境，當你在本地端開發完成後，可以無痛部屬到伺服器 / 雲端，而初期會遇到以下 Docker 相關名詞，本書將一一介紹。

Image

映像檔，如果有用過 VM 虛擬機，那應該很熟悉這名詞，Docker 一切的 Container，都是根據 Image 所建立的，如果要建立資料庫等服務，只需要在 Docker Hub 上，下載由其他人建好的 Image，就能直接啟動相關服務，Image 可以看成 Docker 最基本的元素，小到單一服務，大到整個架構，都是由最基礎的 Image 堆疊出來的。

Docker Hub

由 Docker 官方所建立的 Image 圖書館、社區，以下引述官方網站介紹

> *Docker Hub is the world's largest library and community for*
> *container images.*

裡面包含了 Ubuntu、MySQL、Mongodb、Python、PostgresSQL、Redis 等各種服務的 Image，在過去，安裝以上服務，都需要不同的安裝方式，到了 Docker 的世界，你只需要學習 Docker 指令，就可以統一應用在架設這些服務上。

Container

容器，是根據 Image 所建立出對應的服務，例如使用 MySQL 的 Image，就可以建立出 MySQL 的 Container，這樣你就有 MySQL 資料庫可以使用了。

Volume

在使用 Docker 啟動資料庫的情境下，希望資料能永久保存，不會因為 Container 突然關閉，造成資料遺失，這時就需要借助 Volume，將 Container 內部指令的空間，與外部實體機做連接，這樣當 Container 需要做更新時，關閉重啟，資料庫的資料依然會存在。

讀者可能會想問，Docker 與 VM 有什麼差別？ VM 比較偏向作業系統層面，而 Docker 偏向應用服務層面。

對於初學者來說，Docker 最大的好處，是簡單的幫你安裝好各種服務，以分散式爬蟲為例，你會需要資料庫 MySQL、分散式工具 RabbitMQ，這些 Docker 都能非常簡易的協助你部屬，而 VM 做不到這點。

對於進階者來說，你可能會接觸到 CICD、Docker Swarm、K8s 等架構，這些都是以 Docker 為基礎下去進行，如測試、上線部屬等。以現代架構下，一定會接觸到分散式，Web、API、Backend 等服務，都是分散在多台機器上，借助 Docker，你可以一鍵更新所有機器上的服務，這也是 VM 做不到的。

3.3　安裝 Docker

以下將分別介紹，Ubuntu 跟 Windows 環境下，Docker 的安裝方法。

Ubuntu

打開 terminal，以下指令皆在 terminal 視窗下執行（也可在 VS Code 下的 terminal 執行）

安裝

```
sudo apt-get update
sudo apt-get install -y docker.io
```

將使用者加入 Docker group

```
sudo usermod -aG docker user_name
```

這時需要先登出再重新登入，或是關閉這個 terminal，重新開新的一個，之後測試 Docker

```
docker ps
```

結果如下圖，代表安裝成功

以上完成了 Linux 上 Docker 的安裝。

Windows

到以下連結下載 Docker

https://docs.docker.com/docker-for-windows/install/

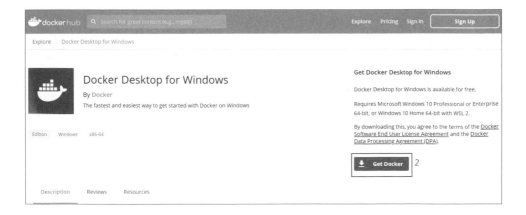

安裝完後，需額外設定 WSL。

目前在 VS Code terminal 視窗下，執行以下指令

```
docker ps
```

會出現以下失敗訊息

```
cannot connect to the docker daemon at unix ///var/run/docker.sock. is
the docker daemon running
```

需要先打開 Docker 設定，勾選以下選項

Expose daemon on tcp://localhost:2375 without tls

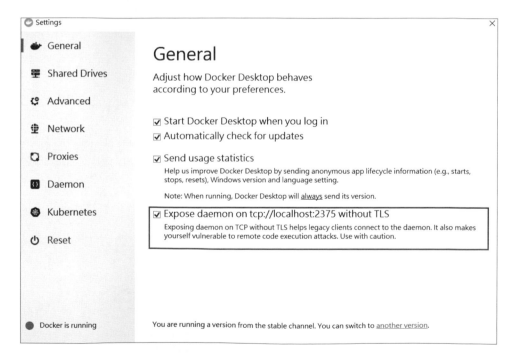

並將以下指令添加到 ~/.bashrc

```
export DOCKER_HOST=127.0.0.1:2375
```

開一個新的 terminal 視窗,並執行

```
docker ps
```

如果結果與下圖相同,代表安裝成功

```
PROBLEMS   OUTPUT   DEBUG CONSOLE   TERMINAL

sam@DESKTOP-IKT69L5:~/FinMindBook$ docker ps
CONTAINER ID   IMAGE     COMMAND     CREATED     STATUS     PORTS     NAMES
sam@DESKTOP-IKT69L5:~/FinMindBook$ █
```

如果是新版 Windows 的讀者,可能會遇到 WSL2 的問題,本書用的是
WSL 1,如果是 WSL 2 的讀者,請參考以下連結,否則在 docker 的部
分,會遇到問題。

https://docs.microsoft.com/zh-tw/windows/wsl/tutorials/wsl-containers

3.4 安裝 Docker-Compose

根據官方文件 https://docs.docker.com/compose/install/,Linux 部分,使用
以下指令安裝

```
sudo curl -L "https://github.com/docker/compose/releases/
download/1.29.1/
docker-compose-$(uname -s)-$(uname -m)" -o /usr/local/bin/docker-compose

sudo chmod +x /usr/local/bin/docker-compose
```

Windows 部分，只要安裝的 Docker Desktop 即可。

有 Docker 之後，本書用到的服務，全部都是一行指令安裝完成。接下來將介紹 ETL 資料工程幾個步驟，資料收集（爬蟲）、資料清理、存入資料庫，但是在這之前，先介紹雲端，以便後續實際應用的案例分享。

Chapter

04

雲端

4.1 為什麼要用雲端？

在 Chapter 5，將介紹分散式爬蟲，而分散式當然不可能只用一台電腦完成，至少需要 2 台電腦做分散，一般人不會有 2 台電腦，這時雲端是個好選擇，本章節介紹一個非常便宜的雲端解決方案，Linode，最低一個月一台雲端機器只要 5 美金，24 小時不斷電、不斷網、不用維護，以自己架設伺服器來說，光網路費就超過 5 美金了。

https://www.linode.com/pricing/

以下擷取 linode 部分內容

RAM	CPU	Storage	Transfer	Network In	Network Out	Price	
1 GB	1 Core	25 GB SSD	1 TB	40 Gbps	1000 Mbps	$5 / mo	($.0075 / hr)

這時讀者不用擔心，是不是馬上要花錢了呢？ GCP、AWS 都有提供免費額度，同樣的，Linode 也有，筆者這裡準備一個 Linode 網址，點進去註冊，就能獲得 100 美金的額度

https://www.linode.com/lp/free-credit-short/

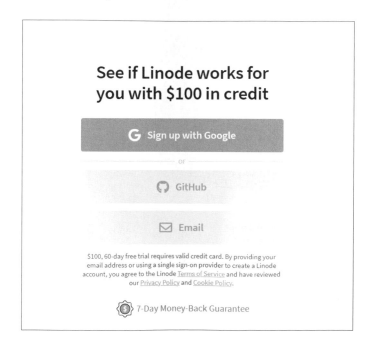

進入連結後，會看到上圖，使用這連結註冊，馬上會獲得 100 美金的額度，足夠讀者同時開 3 台機器（一台 5 美金，三台 15 美金），運行半年以上的時間。

另外的重點是，Linode 每台都提供**固定 IP**，2 台機器就有 2 個 IP，搭配 Python 的 sleep 函數，不用擔心爬蟲被 ban IP 了，也不用擔心用 sleep 效能差，多開幾台就解決了。

下一章節將介紹分散式爬蟲，初期可以開兩台機器嘗試，300 台幣 / 月（10 美金 / 月），Linode 也可以根據使用的小時（0.0075/hr）收費，做完本書範例後，不想使用，也可以刪除，就不用付到一個月的錢了。

資料收集

5.1 Python 環境設置

5.1.1 為什麼需要設置環境？

在開始介紹爬蟲之前，需要先設置 Python 開發環境。

為什麼需要設置開發環境呢？一般開發上，不會只寫一套程式，可能開發多個專案、可能 clone 別人 GitHub 上的專案下來執行，因此，獨立的 Python 環境，有助於安裝 Python Package 時，不會影響其他專案、造成套件衝突等等。

5.1.2　Python 環境、套件管理工具 -- pipenv

本書中，將介紹 pipenv，pipenv 是 Python 官方推薦的套件管理工具，
https://packaging.python.org/guides/tool-recommendations/

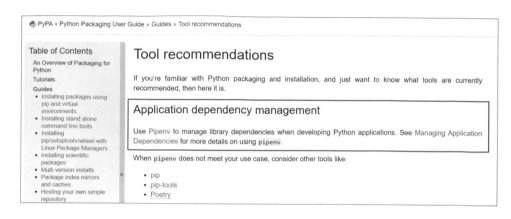

當然讀者也可以使用 conda、Virtualenv 等工具，pipenv 與前面幾個最大
的差別在於，會自動幫你建立 Python 虛擬環境，並自動管理 Python 使用
的 Package，自動 vs 手動，當然自動會比較好。

❏　安裝方式

首先，需要先安裝 pyenv

```
apt-get update
apt install curl git bzip2 -y
curl https://pyenv.run | bash
```

環境設定

```
echo 'export LC_ALL=C.UTF-8' >> ~/.bashrc
echo 'export LANG=C.UTF-8' >> ~/.bashrc
echo 'export PYENV_ROOT="$HOME/.pyenv"' >> ~/.bashrc
echo 'export PATH="$PYENV_ROOT/shims:$PATH"' >> ~/.bashrc
echo 'export PATH="$PYENV_ROOT/bin:$PATH"' >> ~/.bashrc
echo -e 'if command -v pyenv 1>/dev/null 2>&1; then\n  eval "$(pyenv
init -)"\nfi' >> ~/.bashrc
exec $SHELL
```

使 用 pyenv 安 裝 miniconda，你 可 自 行 選 擇 版 本 安 裝，這 邊 選 擇
miniconda 3-4.3.30

```
pyenv install miniconda3-4.3.30
```

設定 global Python 環境

```
pyenv global miniconda3-4.3.30
```

安裝 pipenv

```
pip install pipenv
```

當你使用 pipenv 時，就自動包含了獨立環境，以下將會介紹使用 pipenv
的方式。首先 pipenv 安裝套件方法與 pip 一樣，以 Flask 為例

```
pipenv install flask
```

註：如果遇到 pipenv 無法找到對應的 Python 路徑時，可以使用以下指
令，強制設定路徑

```
pipenv --python ~/.pyenv/versions/miniconda3-4.3.30/bin/python
```

安裝成功畫面如下

```
PROBLEMS    OUTPUT    DEBUG CONSOLE    TERMINAL

sam@DESKTOP-IKT69L5:~/FinMindBook$ pipenv install flask
Creating a Pipfile for this project…
Installing flask…
Adding flask to Pipfile's [packages]…
✓ Installation Succeeded
Pipfile.lock not found, creating…
Locking [dev-packages] dependencies…
Locking [packages] dependencies…
Building requirements...
Resolving dependencies...
✓ Success!
Updated Pipfile.lock (e239e5)!
Installing dependencies from Pipfile.lock (e239e5)…
🐍                              0/0 — 00:00:00
To activate this project's virtualenv, run pipenv shell.
Alternatively, run a command inside the virtualenv with pipenv run.
sam@DESKTOP-IKT69L5:~/FinMindBook$ ▮
```

資料夾下，會出現兩個檔案，

Pipfile

```
  File  Edit  Selection  View  Go  Debug  Terminal  Help        Pipfile - book [WSL: Ubuntu-18.04] - Visual St

  EXPLORER                                  ≡ Pipfile        ×

  ∨ OPEN EDITORS                            ≡ Pipfile
     ×  ≡ Pipfile                            1    [[source]]
  ∨ BOOK [WSL: UBUNTU-18.04]                 2    name = "pypi"
     ≡ Pipfile                               3    url = "https://pypi.org/simple"
     {} Pipfile.lock          1              4    verify_ssl = true
                                             5
                                             6    [dev-packages]
                                             7
                                             8    [packages]
                                             9    flask = "*"           2
                                            10
                                            11    [requires]
                                            12    python_version = "3.6"
                                            13
```

Pipfile.lock

```
≡ Pipfile          {} Pipfile.lock ×
{} Pipfile.lock > {} default > {} importlib-metadata > ⌨ markers
20              "hashes": [
21                  "sha256:8c04c11192119b1ef78ea049e0a6f0463e4c48ef00a30160c704337586f3ad7a",
22                  "sha256:fba402a4a47334742d782209a7c79bc448911afe1149d07bdabdf480b3e2f4b6"
23              ],
24              "markers": "python_version >= '3.6'",
25              "version": "==8.0.1"
26          },
27          "dataclasses": {
28              "hashes": [
29                  "sha256:0201d89fa866f68c8ebd9d08ee6ff50c0b255f8ec63a71c16fda7af82bb887bf",
30                  "sha256:8479067f342acf957dc82ec415d355ab5edb7e7646b90dc6e2fd1d96ad084c97"
31              ],
32              "markers": "python_version < '3.7'",
33              "version": "==0.8"
34          },
35          "flask": {
36              "hashes": [
37                  "sha256:1c4c257b1892aec1398784c63791cbaa43062f1f7aeb555c4da961b20ee68f55",
38                  "sha256:a6209ca15eb63fc9385f38e452704113d679511d9574d09b2cf9183ae7d20dc9"
39              ],
40              "index": "pypi",
41              "version": "==2.0.1"
42          },
```

在 Pipfile 中，packages 底下的 Flask，是剛剛安裝的 package，pipenv 會幫你管理所使用的套件，requires 底下的 python_version 是目前使用的版本。

Pipfile.lock 中，有 flask package 實際使用的版本號與 hashes。

接著，單純執行 Python 並 import flask，會出現

ModuleNotFoundError: No module named 'flask'

```
sam@DESKTOP-IKT69L5:~/FinMindBook$ python
Python 3.6.3 |Anaconda, Inc.| (default, Oct 13 2017, 12:02:49)
[GCC 7.2.0] on linux
Type "help", "copyright", "credits" or "license" for more information.
im>>> import flask
Traceback (most recent call last):
  File "<stdin>", line 1, in <module>
ModuleNotFoundError: No module named 'flask'
>>>
```

這時你可能會有疑問，上面不是安裝了嗎？為什麼沒有 package，這就是 pipenv 的其中一個特點，自動建立獨立環境，剛剛安裝 flask 時，是安裝到 pipenv 建立的獨立環境中，不會安裝到系統環境，所以這時 import 才會失敗。那要如何啟動 pipenv 獨立環境呢？以下指令即可開啟獨立環境

```
pipenv run python
```

pipenv 進入獨立環境、並啟動 Python，兩個動作包在一起，接著再執行，import flask，這時就正常了

```
sam@DESKTOP-IKT69L5:~/FinMindBook$ pipenv run python
Creating a Pipfile for this project…
Python 3.6.9 (default, Jan 26 2021, 15:33:00)
[GCC 8.4.0] on linux
Type "help", "copyright", "credits" or "license" for more information.
>>> import flask
>>>
```

以上，就完成了 Python 環境設置。接下來將介紹爬蟲、資料庫安裝、分散式。

5.2 | 爬蟲

5.2.1 什麼是爬蟲？

爬蟲是使用程式，自動化的方式，擷取特定網站資訊，收集資料的方式。而一般網站，為了防範程式暴力爬蟲，可能會設置驗證碼，或是一定時間內大量連續訪問，會禁止你連線一段時間，以下本書將以證交所、櫃買中心和期交所爬蟲為範例。

程式架構如下，Pipfile 就是上面介紹的套件管理工具，src 底下的 taifex_crawler.py、tpex_crawler.py、twse_crawler.py 分別是期交所、櫃買中心、證交所爬蟲。

https://github.com/FinMind/FinMindBook/tree/master/DataEngineering/Chapter5/5.2

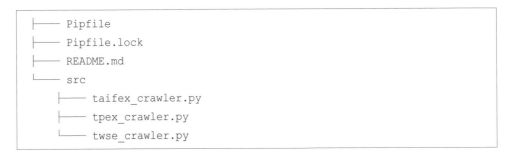

```
├──── Pipfile
├──── Pipfile.lock
├──── README.md
└──── src
    ├──── taifex_crawler.py
    ├──── tpex_crawler.py
    └──── twse_crawler.py
```

5.2.2 證交所爬蟲範例

https://www.twse.com.tw/zh/page/trading/exchange/MI_INDEX.html

證交所網址如上，以股價為例，在做爬蟲之前，需要觀察在網頁上操作的過程中，對網頁做了什麼樣的動作，並用程式模仿這段動作，就可以取得想要的資訊。

1. 在網站上點擊右鍵，選取**檢查**，檢查可以幫助觀察，當查詢股價時，實際上發什麼 request 請求給證交所伺服器，之後再用程式去模擬這段 request

2. 選取 **Network**，Network 包含了接下來對伺服器發送的 request 與 response，讀者需要從這觀察，請求 request 與回應 response

3. 選擇全部，這是要爬取的股價分類，

4. 點選查詢之後，右邊 Network 部分，會出現伺服器的回應，選取第一個 **MI_INDEX?response=json**，觀察查詢動作，是針對哪個網址發送 request，發送內容是什麼，收到的 response 是不是股價資訊要的

5. 點選 **Headers**，觀察到查詢這個動作，是對

   https://www.twse.com.tw/exchangeReport/MI_INDEX?response=json&date=20210309&type=ALL

   發送 request 請求。

   這裡的 **GET**，是 request 的方式，有些網站是 **POST**，這要特別注意

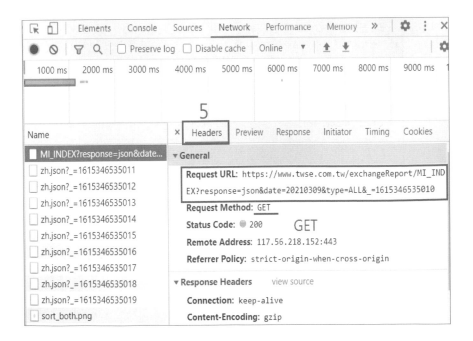

6. 觀察 header，待會書中下面的 Python Code 會用到，在模擬瀏覽器登入時所帶的參數

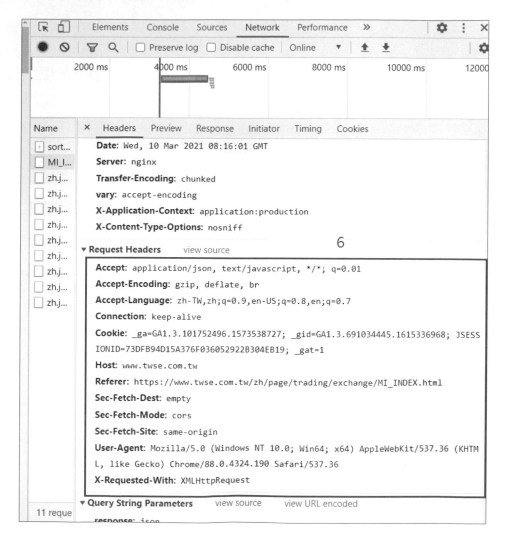

7. 點選 **Preview**，這裡可以看到對網站發送請求時得到的 data，這裡有
data1 到 data9，那怎麼判斷，哪些是股價資訊呢？

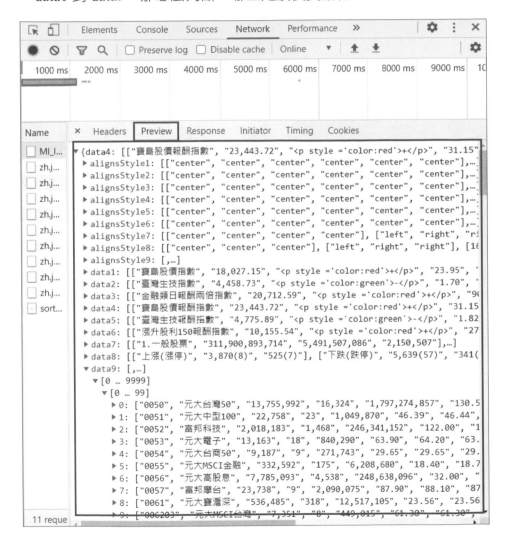

8. 把網頁滾輪滑到最下面，到**每日收盤行情**的位置，接著來比對一下
   Preview，觀察出，**data9** 是所需要的股價資料，欄位名稱在 **fields9**，
   找到 data 後，接下來開始爬蟲。

					110年03月09日 每日收盤行情(全部)										

總頁 10 ∨ 筆

證券代號	證券名稱	成交股數	成交筆數	成交金額	開盤價	最高價	最低價	收盤價	漲跌(+/-)	漲跌價差	最後揭示買價	最後揭示買量	最後揭示賣價	最後揭示賣量	本益比
0050	元大台灣50	13,755,992	16,324	1,797,274,857	130.55	131.60	129.75	131.35	-	0.50	131.35	9	131.40	16	0.00
0051	元大中型100	22,758	23	1,049,870	46.39	46.44	45.74	46.20	-	0.19	46.35	3	46.36	1	0.00
0052	富邦科技	2,018,183	1,468	246,341,152	122.00	123.15	121.30	123.00	-	0.85	122.95	1	123.00	9	0.00
0053	元大電子	13,163	18	840,290	63.90	64.20	63.60	64.00	-	0.30	63.90	1	64.30	10	0.00
0054	元大台商50	9,187	9	271,743	29.65	29.65	29.40	29.60	-	0.45	29.54	55	29.70	1	0.00
0055	元大MSCI金融	332,592	175	6,208,680	18.40	18.77	18.40	18.75	+	0.07	18.72	1	18.75	8	0.00
0056	元大高股息	7,785,093	4,538	248,638,096	32.00	32.01	31.84	32.00	-	0.00	31.99	1	32.00	259	0.00
0057	富邦摩台	23,738	9	2,090,075	87.90	88.10	87.85	88.10	-	0.55	88.05	10	88.50	10	0.00
0061	元大寶滬深	536,485	318	12,517,105	23.56	23.56	22.95	23.42	-	0.29	23.41	6	23.42	49	0.00
006203	元大MSCI台灣	7,351	8	449,015	61.30	61.30	60.80	61.30	-	0.30	61.10	3	61.35	35	0.00

上頁 1 2 3 4 5 … 2149 下頁

註：證交所早期資料跟近年格式不同，以股價來說，2009 以前的資料，
股價在 response 中的 data8，2009 以後的資料在 data9，因此在做爬蟲
時，需要多看幾筆資料來判斷，以下的程式碼會處理這種情境。

9. 爬蟲程式碼，可參考以下連結：

https://github.com/FinMind/FinMindBook/blob/master/DataEngineering/
Chapter5/5.2/src/twse_crawler.py

在開始之前，你需要先執行以下指令，安裝 Package

```
pipenv install pandas requests pydantic loguru
```

爬蟲程式碼：

```
src/twse_crawler.py
import datetime
import sys
import time
import typing

import pandas as pd
import requests
from loguru import logger
from pydantic import BaseModel


def clear_data(
    df: pd.DataFrame,
) -> pd.DataFrame:
    """資料清理，將文字轉成數字"""
    df["Dir"] = (
        df["Dir"]
        .str.split(">")
        .str[1]
        .str.split("<")
        .str[0]
    )
    df["Change"] = (
        df["Dir"] + df["Change"]
    )
    df["Change"] = (
        df["Change"]
        .str.replace(" ", "")
        .str.replace("X", "")
```

```python
            .astype(float)
    )
    df = df.fillna("")
    df = df.drop(["Dir"], axis=1)
    for col in [
        "TradeVolume",
        "Transaction",
        "TradeValue",
        "Open",
        "Max",
        "Min",
        "Close",
        "Change",
    ]:
        df[col] = (
            df[col]
            .astype(str)
            .str.replace(",", "")
            .str.replace("X", "")
            .str.replace("+", "")
            .str.replace("----", "0")
            .str.replace("---", "0")
            .str.replace("--", "0")
        )
    return df


def colname_zh2en(
    df: pd.DataFrame,
    colname: typing.List[str],
) -> pd.DataFrame:
    """資料欄位轉換，英文有助於接下來存入資料庫"""
```

```python
    taiwan_stock_price = {
        "證券代號": "StockID",
        "證券名稱": "",
        "成交股數": "TradeVolume",
        "成交筆數": "Transaction",
        "成交金額": "TradeValue",
        "開盤價": "Open",
        "最高價": "Max",
        "最低價": "Min",
        "收盤價": "Close",
        "漲跌(+/-)": "Dir",
        "漲跌價差": "Change",
        "最後揭示買價": "",
        "最後揭示買量": "",
        "最後揭示賣價": "",
        "最後揭示賣量": "",
        "本益比": "",
    }
    df.columns = [
        taiwan_stock_price[col]
        for col in colname
    ]
    df = df.drop([""], axis=1)
    return df


def twse_header():
    """網頁瀏覽時，所帶的 request header 參數，模仿瀏覽器發送 request"""
    return {
        "Accept": "application/json, text/javascript, */*; q=0.01",
        "Accept-Encoding": "gzip, deflate",
        "Accept-Language": "zh-TW,zh;q=0.9,en-US;q=0.8,en;q=0.7",
```

```
        "Connection": "keep-alive",
        "Host": "www.twse.com.tw",
        "Referer": "https://www.twse.com.tw/zh/page/trading/exchange/
MI_INDEX.html",
        "User-Agent": "Mozilla/5.0 (Windows NT 10.0; Win64;
x64) AppleWebKit/537.36 (KHTML, like Gecko) Chrome/71.0.3578.98
Safari/537.36",
        "X-Requested-With": "XMLHttpRequest",
    }


def crawler_twse(
    date: str,
) -> pd.DataFrame:
    """
    證交所網址
    https://www.twse.com.tw/zh/page/trading/exchange/MI_INDEX.html
    """
    # headers 中的 Request url
    url = (
        "https://www.twse.com.tw/exchangeReport/MI_INDEX"
        "?response=json&date={date}&type=ALL"
    )
    url = url.format(
        date=date.replace("-", "")
    )
    # 避免被證交所 ban ip, 在每次爬蟲時, 先 sleep 5 秒
    time.sleep(5)
    # request method
    res = requests.get(
        url, headers=twse_header()
    )
```

```python
    if (
        res.json()["stat"]
        == "很抱歉，沒有符合條件的資料!"
    ):
        # 如果 date 是周末，會回傳很抱歉，沒有符合條件的資料!
        return pd.DataFrame()
    # 2009 年以後的資料，股價在 response 中的 data9
    # 2009 年以後的資料，股價在 response 中的 data8
    # 不同格式，在證交所的資料中，是很常見的，
    # 沒資料的情境也要考慮進去，例如現在週六沒有交易，但在 2007 年週六是有
交易的
    try:
        if "data9" in res.json():
            df = pd.DataFrame(
                res.json()["data9"]
            )
            colname = res.json()[
                "fields9"
            ]
        elif "data8" in res.json():
            df = pd.DataFrame(
                res.json()["data8"]
            )
            colname = res.json()[
                "fields8"
            ]
        elif res.json()["stat"] in [
            "查詢日期小於93年2月11日，請重新查詢!",
            "很抱歉，沒有符合條件的資料!",
        ]:
            return pd.DataFrame()
    except BaseException:
```

```
            return pd.DataFrame()

    if len(df) == 0:
        return pd.DataFrame()
    # 欄位中英轉換
    df = colname_zh2en(
        df.copy(), colname
    )
    df["date"] = date
    return df


class TaiwanStockPrice(BaseModel):
    StockID: str
    TradeVolume: int
    Transaction: int
    TradeValue: int
    Open: float
    Max: float
    Min: float
    Close: float
    Change: float
    date: str


def check_schema(
    df: pd.DataFrame,
) -> pd.DataFrame:
    """檢查資料型態, 確保每次要上傳資料庫前, 型態正確"""
    df_dict = df.to_dict("records")
    df_schema = [
        TaiwanStockPrice(**dd).__dict__
```

```
        for dd in df_dict
    ]
    df = pd.DataFrame(df_schema)
    return df


def gen_date_list(
    start_date: str, end_date: str
) -> typing.List[str]:
    """建立時間列表, 用於爬取所有資料"""
    start_date = (
        datetime.datetime.strptime(
            start_date, "%Y-%m-%d"
        ).date()
    )
    end_date = (
        datetime.datetime.strptime(
            end_date, "%Y-%m-%d"
        ).date()
    )
    days = (
        end_date - start_date
    ).days + 1
    date_list = [
        str(
            start_date
            + datetime.timedelta(
                days=day
            )
        )
        for day in range(days)
    ]
```

```python
        return date_list



def main(
    start_date: str, end_date: str
):
    """證交所寫明，※ 本資訊自民國93年2月11日起提供"""
    date_list = gen_date_list(
        start_date, end_date
    )
    for date in date_list:
        logger.info(date)
        df = crawler_twse(date)
        if len(df) > 0:
            # 資料清理
            df = clear_data(df.copy())
            # 檢查資料型態
            df = check_schema(df.copy())
            # 這邊先暫時存成 file，下個章節將會上傳資料庫
            df.to_csv(
                f"taiwan_stock_price_twse_{date}.csv",
                index=False,
            )



if __name__ == "__main__":
    start_date, end_date = sys.argv[1:]
    main(start_date, end_date)
```

執行以下指令，即可爬取 2021-03-20 到 2021-03-25 的證交所股價。

```
pipenv run python src/twse_crawler.py 2021-03-20 2021-03-25
```

基本上證交所其他資料的格式,如三大法人、融資融券,都與股價相同,因此模仿上面的程式碼,都可以正常的爬取資料,但證交所只有上市資料,而平常股市分析,會使用上市與上櫃資料,接下來,介紹櫃買中心爬蟲。

5.2.3 櫃買中心爬蟲範例

https://www.tpex.org.tw/web/stock/aftertrading/otc_quotes_no1430/stk_wn1430.php?l=zh-tw

櫃買中心網址如上,爬蟲方式,基本上跟證交所一樣,就不一步步贅述:

爬蟲程式碼,可參考以下連結:

https://github.com/FinMind/FinMindBook/blob/master/DataEngineering/Chapter5/5.2/src/tpex_crawler.py

```
src/tpex_crawler.py
import datetime
import sys
import time
import typing

import pandas as pd
import requests
from loguru import logger
from pydantic import BaseModel


def clear_data(
    df: pd.DataFrame,
) -> pd.DataFrame:
    """資料清理, 將文字轉成數字"""
    for col in [
        "TradeVolume",
        "Transaction",
        "TradeValue",
        "Open",
        "Max",
        "Min",
        "Close",
        "Change",
    ]:
        df[col] = (
            df[col]
            .astype(str)
            .str.replace(",", "")
            .str.replace("X", "")
            .str.replace("+", "")
            .str.replace("----", "0")
            .str.replace("---", "0")
```

```
            .str.replace("--", "0")
            .str.replace(" ", "")
            .str.replace("除權息", "0")
            .str.replace("除息", "0")
            .str.replace("除權", "0")
        )
    return df


def set_column(
    df: pd.DataFrame,
) -> pd.DataFrame:
    """設定資料欄位名稱"""
    df.columns = [
        "StockID",
        "Close",
        "Change",
        "Open",
        "Max",
        "Min",
        "TradeVolume",
        "TradeValue",
        "Transaction",
    ]
    return df


def tpex_header():
    """網頁瀏覽時，所帶的 request header 參數，模仿瀏覽器發送 request"""
    return {
        "Accept": "application/json, text/javascript, */*; q=0.01",
        "Accept-Encoding": "gzip, deflate",
        "Accept-Language": "zh-TW,zh;q=0.9,en-US;q=0.8,en;q=0.7",
        "Connection": "keep-alive",
```

```
        "Host": "www.tpex.org.tw",
        "Referer": "https://www.tpex.org.tw/web/stock/aftertrading/
otc_quotes_no1430/stk_wn1430.php?l=zh-tw",
        "User-Agent": "Mozilla/5.0 (Windows NT 10.0; Win64; x64)
AppleWebKit/537.36 (KHTML, like Gecko) Chrome/73.0.3683.103
Safari/537.36",
        "X-Requested-With": "XMLHttpRequest",
    }


def convert_date(date: str) -> str:
    year, month, day = date.split("-")
    year = int(year) - 1911
    return f"{year}/{month}/{day}"


def crawler_tpex(
    date: str,
) -> pd.DataFrame:
    """
    櫃買中心網址
    https://www.tpex.org.tw/web/stock/aftertrading/otc_quotes_no1430/
stk_wn1430.php?l=zh-tw
    """
    # headers 中的 Request url
    url = (
        "https://www.tpex.org.tw/web/stock/aftertrading/"
        "otc_quotes_no1430/stk_wn1430_result.php?"
        "l=zh-tw&d={date}&se=AL"
    )
    url = url.format(
        date=convert_date(date)
    )
    # 避免被櫃買中心 ban ip, 在每次爬蟲時, 先 sleep 5 秒
```

```
    time.sleep(5)
    # request method
    res = requests.get(
        url, headers=tpex_header()
    )
    data = res.json().get("aaData", "")
    if not data:
        return pd.DataFrame()
    df = pd.DataFrame(data)

    if len(df) == 0:
        return pd.DataFrame()
    # 櫃買中心回傳的資料, 並無資料欄位, 因此這裡直接用 index 取特定欄位
    df = df[[0, 2, 3, 4, 5, 6, 7, 8, 9]]
    # 欄位中英轉換
    df = set_column(df.copy())
    df["date"] = date
    return df


class TaiwanStockPrice(BaseModel):
    StockID: str
    TradeVolume: int
    Transaction: int
    TradeValue: int
    Open: float
    Max: float
    Min: float
    Close: float
    Change: float
    date: str


def check_schema(
```

```
    df: pd.DataFrame,
) -> pd.DataFrame:
    """檢查資料型態，確保每次要上傳資料庫前，型態正確"""
    df_dict = df.to_dict("records")
    df_schema = [
        TaiwanStockPrice(**dd).__dict__
        for dd in df_dict
    ]
    df = pd.DataFrame(df_schema)
    return df


def gen_date_list(
    start_date: str, end_date: str
) -> typing.List[str]:
    """建立時間列表，用於爬取所有資料"""
    start_date = (
        datetime.datetime.strptime(
            start_date, "%Y-%m-%d"
        ).date()
    )
    end_date = (
        datetime.datetime.strptime(
            end_date, "%Y-%m-%d"
        ).date()
    )
    days = (
        end_date - start_date
    ).days + 1
    date_list = [
        str(
            start_date
            + datetime.timedelta(
                days=day
```

```
            )
        )
        for day in range(days)
    ]
    return date_list


def main(
    start_date: str, end_date: str
):
    """櫃買中心寫明，本資訊自民國96年7月起開始提供"""
    date_list = gen_date_list(
        start_date, end_date
    )
    for date in date_list:
        logger.info(date)
        df = crawler_tpex(date)
        if len(df) > 0:
            # 資料清理
            df = clear_data(df.copy())
            # 檢查資料型態
            df = check_schema(df.copy())
            # 這邊先暫時存成 file，下個章節將會上傳資料庫
            df.to_csv(
                f"taiwan_stock_price_tpex_{date}.csv",
                index=False,
            )


if __name__ == "__main__":
    start_date, end_date = sys.argv[1:]
    main(start_date, end_date)
```

執行以下指令，即可爬取 2021-03-20 到 2021-03-25 的櫃買中心股價。

```
pipenv run python src/tpex_crawler.py 2021-03-20 2021-03-25
```

與證交所相同，櫃買中心的其他資料格式也一致，本書到此，對於證交所、櫃買中心的爬蟲，可以說是掌握八成了，以下將介紹期交所爬蟲。

5.2.4　期交所爬蟲範例

https://www.taifex.com.tw/cht/3/futDailyMarketView

期交所網址如上，跟證交所、櫃買中心一樣的方法，另外期貨與選擇權相同，這邊以期貨為例：

這邊稍微不同，是用 POST，需要將右邊的滑輪，拉到最下面，觀察 POST 發送什麼 Form Data。

接著以下的程式碼，會模仿瀏覽器，發送 POST 給期交所伺服器。

可參考以下連結：

https://github.com/FinMind/FinMindBook/blob/master/DataEngineering/
Chapter5/5.2/src/taifex_crawler.py

```
src/taifex_crawler.py
import datetime
import io
import sys
import time
import typing


import pandas as pd
import requests
```

```python
from loguru import logger
from pydantic import BaseModel


def futures_header():
    """網頁瀏覽時，所帶的 request header 參數，模仿瀏覽器發送 request"""
    return {
        "Accept": "text/html,application/xhtml+xml,application/xml;
q=0.9,image/webp,image/apng,*/*;q=0.8,application/signed-exchange;
v=b3;q=0.9",
        "Accept-Encoding": "gzip, deflate, br",
        "Accept-Language": "zh-TW,zh;q=0.9,en-US;q=0.8,en;q=0.7",
        "Cache-Control": "no-cache",
        "Connection": "keep-alive",
        "Content-Length": "101",
        "Content-Type": "application/x-www-form-urlencoded",
        "Host": "www.taifex.com.tw",
        "Origin": "https://www.taifex.com.tw",
        "Pragma": "no-cache",
        "Referer": "https://www.taifex.com.tw/cht/3/dlFutDailyMarketView",
        "Sec-Fetch-Dest": "document",
        "Sec-Fetch-Mode": "navigate",
        "Sec-Fetch-Site": "same-origin",
        "Sec-Fetch-User": "?1",
        "Upgrade-Insecure-Requests": "1",
        "User-Agent": "Mozilla/5.0 (Windows NT 10.0; Win64; x64)
AppleWebKit/537.36 (KHTML, like Gecko) Chrome/81.0.4044.113 Safari/537.36",
    }


def colname_zh2en(
    df: pd.DataFrame,
```

```python
) -> pd.DataFrame:
    """資料欄位轉換, 英文有助於接下來存入資料庫"""
    colname_dict = {
        "交易日期": "date",
        "契約": "FuturesID",
        "到期月份(週別)": "ContractDate",
        "開盤價": "Open",
        "最高價": "Max",
        "最低價": "Min",
        "收盤價": "Close",
        "漲跌價": "Change",
        "漲跌%": "ChangePer",
        "成交量": "Volume",
        "結算價": "SettlementPrice",
        "未沖銷契約數": "OpenInterest",
        "交易時段": "TradingSession",
    }
    df = df.drop(
        [
            "最後最佳買價",
            "最後最佳賣價",
            "歷史最高價",
            "歷史最低價",
            "是否因訊息面暫停交易",
            "價差對單式委託成交量",
        ],
        axis=1,
    )
    df.columns = [
        colname_dict[col]
        for col in df.columns
    ]
```

```python
    return df


def clean_data(
    df: pd.DataFrame,
) -> pd.DataFrame:
    """資料清理"""
    df["date"] = df["date"].str.replace(
        "/", "-"
    )
    df["ChangePer"] = df[
        "ChangePer"
    ].str.replace("%", "")
    df["ContractDate"] = (
        df["ContractDate"]
        .astype(str)
        .str.replace(" ", "")
    )
    if "TradingSession" in df.columns:
        df["TradingSession"] = df[
            "TradingSession"
        ].map(
            {
                "一般": "Position",
                "盤後": "AfterMarket",
            }
        )
    else:
        df[
            "TradingSession"
        ] = "Position"
    for col in [
```

```
        "Open",
        "Max",
        "Min",
        "Close",
        "Change",
        "ChangePer",
        "Volume",
        "SettlementPrice",
        "OpenInterest",
    ]:
        df[col] = (
            df[col]
            .replace("-", "0")
            .astype(float)
        )
    df = df.fillna(0)
    return df


def crawler_futures(
    date: str,
) -> pd.DataFrame:
    """期交所爬蟲"""
    url = "https://www.taifex.com.tw/cht/3/futDataDown"
    form_data = {
        "down_type": "1",
        "commodity_id": "all",
        "queryStartDate": date.replace(
            "-", "/"
        ),
        "queryEndDate": date.replace(
            "-", "/"
```

```
        ),
    }
    # 避免被期交所 ban ip，在每次爬蟲時，先 sleep 5 秒
    time.sleep(5)
    resp = requests.post(
        url,
        headers=futures_header(),
        data=form_data,
    )
    if resp.ok:
        if resp.content:
            df = pd.read_csv(
                io.StringIO(
                    resp.content.decode(
                        "big5"
                    )
                ),
                index_col=False,
            )
    else:
        return pd.DataFrame()
    return df


class TaiwanFuturesDaily(BaseModel):
    date: str
    FuturesID: str
    ContractDate: str
    Open: float
    Max: float
    Min: float
    Close: float
```

```python
    Change: float
    ChangePer: float
    Volume: float
    SettlementPrice: float
    OpenInterest: int
    TradingSession: str


def check_schema(
    df: pd.DataFrame,
) -> pd.DataFrame:
    """檢查資料型態，確保每次要上傳資料庫前，型態正確"""
    df_dict = df.to_dict("records")
    df_schema = [
        TaiwanFuturesDaily(
            **dd
        ).__dict__
        for dd in df_dict
    ]
    df = pd.DataFrame(df_schema)
    return df


def gen_date_list(
    start_date: str, end_date: str
) -> typing.List[str]:
    """建立時間列表，用於爬取所有資料"""
    start_date = (
        datetime.datetime.strptime(
            start_date, "%Y-%m-%d"
        ).date()
    )
```

```python
    end_date = (
        datetime.datetime.strptime(
            end_date, "%Y-%m-%d"
        ).date()
    )
    days = (
        end_date - start_date
    ).days + 1
    date_list = [
        str(
            start_date
            + datetime.timedelta(
                days=day
            )
        )
        for day in range(days)
    ]
    return date_list


def main(
    start_date: str, end_date: str
):
    date_list = gen_date_list(
        start_date, end_date
    )
    for date in date_list:
        logger.info(date)
        df = crawler_futures(date)
        if len(df) > 0:
            # 欄位中英轉換
            df = colname_zh2en(
```

```
            df.copy()
        )
        # 資料清理
        df = clean_data(df.copy())
        # 檢查資料型態
        df = check_schema(df.copy())
        # 這邊先暫時存成 file，下個章節將會上傳資料庫
        df.to_csv(
            f"taiwan_futures_price_{date}.csv",
            index=False,
        )


if __name__ == "__main__":
    start_date, end_date = sys.argv[1:]
    main(start_date, end_date)
```

執行以下指令，即可爬取 2021-03-20 到 2021-03-25 的期交所期貨價格。

```
pipenv run python src/taifex_crawler.py 2021-03-20 2021-03-25
```

以上，台股資料，大多都能成功爬蟲了，當然爬蟲難度不高，問題是如何管理大量的爬蟲？如果將證交所、櫃買中心、期交所大多的資料都爬下來，可能有 50 隻爬蟲程式，那要如何管理？如果都存成 csv，那這些檔案要如何管理？本書將會在 5.3、5.4 章節，介紹爬蟲後上傳資料庫，在 5.5 章節，介紹 Celery 分散式爬蟲。在這之前，先架設資料庫，並將以上爬蟲資料存入資料庫。

5.3 資料庫架設

5.3.1 什麼是資料庫？

為什麼需要資料庫？可能剛開始接觸爬蟲的工程師，不太清楚資料庫的用途，這時本書用爬蟲，作為資料庫應用的場景。

資料庫是協助儲存、管理各種資料的工具，以股價為例，一年約股市交易約 200 天，當你爬蟲收集十年的歷史股價時，不太可能直接存成 Excel，即使你真的把全部股價存成一個 Excel，那除了股價，還有三大法人、融資券、期貨、選擇權等等的資料，直接存成 Excel 拓展性低，用資料庫來管理，會比較有效率，以下將會介紹如何用 Docker 一鍵架設資料庫。

5.3.2 關聯式資料庫 --- MySQL

在介紹之前，先來安裝 MySQL，將以下程式碼，存成 mysql.yml，或是可以從以下連結下載

https://github.com/FinMind/FinMindBook/blob/master/DataEngineering/Chapter5/5.3/mysql.yml

```
mysql.yml
version: '3.3'
services:

  mysql:
    image: mysql:8.0
    # 設定 mysql 使用原生認證的密碼 hash
    command: mysqld --default-authentication-plugin=mysql_native_password
    ports:
```

```
        # docker publish port 3306 to 3306
        # (將 docker 內部 ip 3306, 跟外部 3306 做連結)
          - 3306:3306
        # - target: 3306
        #   published: 3306
        #   mode: host
    environment: # 環境變數, 設置 db, user, password, root_password
        MYSQL_DATABASE: mydb
        MYSQL_USER: user
        MYSQL_PASSWORD: test
        MYSQL_ROOT_PASSWORD: test
    volumes:
        - mysql:/var/lib/mysql
        # share volumes of docker container to outside volume,
        # let data persist
        # 將 docker container 的 volumes 跟外部 volumes 做連結,
        # 讓資料不會因為, container close or update, 而遺失
    networks:
        - dev


phpmyadmin:
    image: phpmyadmin/phpmyadmin:5.1.0
    links:
        - mysql:db
    ports:
        - 8000:80
    depends_on:
      - mysql
    networks:
        - dev

networks:
```

```
    dev:

volumes:
  mysql:
    external: true
```

先建立 volume，使用以下指令：

```
docker volume create mysql
```

volume 只要建立一次就好，這是確保資料持久性，不會因為 Docker 關閉
而資料遺失，接著使用以下指令，就架好 MySQL 了。

```
docker-compose -f mysql.yml up
```

這時 Docker 會自動去網路上，下載 Image，如下圖：

```
sam@DESKTOP-IKT69L5:~/FinMindBook/DataEngineering/Chapter5/5.3$ docker-compose -f mysql.yml up
Pulling mysql (mysql:8.0)...
8.0: Pulling from library/mysql
b380bbd43752: Pull complete
f23cbf2ecc5d: Pull complete
30cfc6c29c0a: Pull complete
b38609286cbe: Pull complete
8211d9e66cd6: Pull complete
2313f9eeca4a: Pull complete
7eb487d00da0: Pull complete
4d7421c8152e: Pull complete
77f3d8811a28: Pull complete
cce755338cba: Pull complete
69b753046b9f: Pull complete
b2e64b0ab53c: Pull complete
Digest: sha256:6d7d4524463fe6e2b893ffc2b89543c81dec7ef82fb2020a1b27606666464d87
Status: Downloaded newer image for mysql:8.0
Pulling phpmyadmin (phpmyadmin/phpmyadmin:5.1.0)...
5.1.0: Pulling from phpmyadmin/phpmyadmin
45b42c59be33: Downloading [=====================================>     ]  20.38MB/27.1MB
a48991d6909c: Download complete
935e2abd2c2c: Downloading [========>                                  ]  12.88MB/76.68MB
61ccf45ccdb9: Waiting
27b5ac70765b: Waiting
5638b69045ba: Waiting
0fdaed064166: Waiting
e932cec09ced: Waiting
```

下載完畢之後，打開瀏覽器，輸入 http://localhost:8000/，帳號密碼是 root/test：

登入後，會看到以下畫面：

這時有 UI 可以操作,就可以開始使用資料庫了,先建立 financialdata 資料庫,方便接下來爬蟲使用,步驟如下。

成功建立後,畫面如下:

再回來看你剛剛的 terminal，會呈現以下樣子，卡住，無法做其他事：

```
PROBLEMS   OUTPUT   DEBUG CONSOLE   TERMINAL                                                    1: python      ∨   +  ⊡ 🗑  ∧  ×
s NT 10.0; Win64; x64) AppleWebKit/537.36 (KHTML, like Gecko) Chrome/91.0.4472.124 Safari/537.36"
phpmyadmin_1  |  172.19.0.1 - - [07/Jul/2021:16:52:14 +0000] "GET /js/vendor/codemirror/addon/hint/show-hint.js?v=5.1.0 HTTP/1.1" 200 5659 "-" "Mozilla/5.0
(Windows NT 10.0; Win64; x64) AppleWebKit/537.36 (KHTML, like Gecko) Chrome/91.0.4472.124 Safari/537.36"
phpmyadmin_1  |  172.19.0.1 - - [07/Jul/2021:16:52:14 +0000] "GET /js/vendor/codemirror/addon/hint/sql-hint.js?v=5.1.0 HTTP/1.1" 200 3097 "-" "Mozilla/5.0 (
Windows NT 10.0; Win64; x64) AppleWebKit/537.36 (KHTML, like Gecko) Chrome/91.0.4472.124 Safari/537.36"
phpmyadmin_1  |  172.19.0.1 - - [07/Jul/2021:16:52:14 +0000] "GET /js/vendor/codemirror/addon/lint/lint.js?v=5.1.0 HTTP/1.1" 200 3069 "-" "Mozilla/5.0 (Wind
ows NT 10.0; Win64; x64) AppleWebKit/537.36 (KHTML, like Gecko) Chrome/91.0.4472.124 Safari/537.36"
phpmyadmin_1  |  172.19.0.1 - - [07/Jul/2021:16:52:14 +0000] "GET /js/dist/codemirror/addon/lint/sql-lint.js?v=5.1.0 HTTP/1.1" 200 810 "-" "Mozilla/5.0 (Win
dows NT 10.0; Win64; x64) AppleWebKit/537.36 (KHTML, like Gecko) Chrome/91.0.4472.124 Safari/537.36"
phpmyadmin_1  |  172.19.0.1 - - [07/Jul/2021:16:52:14 +0000] "GET /js/vendor/codemirror/lib/codemirror.js?v=5.1.0 HTTP/1.1" 200 106200 "-" "Mozilla/5.0 (Win
dows NT 10.0; Win64; x64) AppleWebKit/537.36 (KHTML, like Gecko) Chrome/91.0.4472.124 Safari/537.36"
phpmyadmin_1  |  172.19.0.1 - - [07/Jul/2021:16:52:14 +0000] "GET /js/dist/console.js?v=5.1.0 HTTP/1.1" 200 10628 "-" "Mozilla/5.0 (Windows NT 10.0; Win64;
x64) AppleWebKit/537.36 (KHTML, like Gecko) Chrome/91.0.4472.124 Safari/537.36"
phpmyadmin_1  |  172.19.0.1 - - [07/Jul/2021:16:52:14 +0000] "GET /themes/pmahomme/img/logo_right.png HTTP/1.1" 200 4592 "-" "Mozilla/5.0 (Windows NT 10.0;
Win64; x64) AppleWebKit/537.36 (KHTML, like Gecko) Chrome/91.0.4472.124 Safari/537.36"
phpmyadmin_1  |  172.19.0.1 - - [07/Jul/2021:16:52:14 +0000] "GET /themes/dot.gif HTTP/1.1" 200 325 "-" "Mozilla/5.0 (Windows NT 10.0; Win64; x64) AppleWebK
it/537.36 (KHTML, like Gecko) Chrome/91.0.4472.124 Safari/537.36"
phpmyadmin_1  |  172.19.0.1 - - [07/Jul/2021:16:52:14 +0000] "GET /themes/pmahomme/css/printview.css?v=5.1.0 HTTP/1.1" 200 954 "-" "Mozilla/5.0 (Windows NT
10.0; Win64; x64) AppleWebKit/537.36 (KHTML, like Gecko) Chrome/91.0.4472.124 Safari/537.36"
phpmyadmin_1  |  172.19.0.1 - - [07/Jul/2021:16:52:14 +0000] "GET /themes/pmahomme/img/b_help.png HTTP/1.1" 200 989 "http://localhost:8000/themes/pmahomme/c
ss/theme.css?v=5.1.0&nocache=4842446188ltr&server=1" "Mozilla/5.0 (Windows NT 10.0; Win64; x64) AppleWebKit/537.36 (KHTML, like Gecko) Chrome/91.0.4472.124
 Safari/537.36"
phpmyadmin_1  |  172.19.0.1 - - [07/Jul/2021:16:52:14 +0000] "GET /favicon.ico HTTP/1.1" 200 22788 "-" "Mozilla/5.0 (Windows NT 10.0; Win64; x64) AppleWebKi
t/537.36 (KHTML, like Gecko) Chrome/91.0.4472.124 Safari/537.36"
```

這時輸入 ctrl+c，即可離開，但如果想要背景執行 mysql 資料庫，怎麼做呢？執行以下指令，在原先指令上，加入 -d，即可背景執行：

```
docker-compose -f mysql.yml up -d
```

呈現以下畫面：

```
sam@DESKTOP-IKT69L5:~/FinMindBook/DataEngineering/Chapter5/5.3$ docker-compose -f mysql.yml up -d
Creating network "53_dev" with the default driver
Creating 53_mysql_1 ... done
Creating 53_phpmyadmin_1 ... done
sam@DESKTOP-IKT69L5:~/FinMindBook/DataEngineering/Chapter5/5.3$
```

就在背景執行了。

執行以下指令，可以看到目前背景正在執行那些 Docker：

```
docker ps
```

```
sam@DESKTOP-IKT69L5:~/FinMindBook/DataEngineering/Chapter5/5.3$ docker ps
CONTAINER ID   IMAGE                          COMMAND               CREATED         STATUS          PORTS
6ab83448cc06   phpmyadmin/phpmyadmin:5.1.0    "/docker-entrypoint..." 14 minutes ago  Up 14 minutes   0.0.0.0:8000->80/tcp, :::800
0->80/tcp                     53_phpmyadmin_1
bc9a660d7186   mysql:8.0                      "docker-entrypoint.s..." 14 minutes ago  Up 14 minutes   0.0.0.0:3306->3306/tcp, :::3
306->3306/tcp, 33060/tcp      53_mysql_1
sam@DESKTOP-IKT69L5:~/FinMindBook/DataEngineering/Chapter5/5.3$
```

剛剛啟動了 mysql 與 phpmyadmin，那要如何關閉呢？執行以下指令就會
關閉 MySQL：

```
docker-compose -f mysql.yml down
```

那什麼是 MySQL 呢？ MySQL 是一種關聯式資料庫，主要用來儲存結構
型資料，優點是免費，算是蠻廣泛被使用的資料庫，而金融資料，大多
也是結構型資料，因此本書使用 MySQL 進行儲存。

由於金融資料是時間序列型資料，MySQL 是關聯式資料庫，並不完全適
合時序資料，如果想要使用專門的時序資料庫，有以下幾種可做選擇，
InfluxDB 、kdb+、TimescaleDB、DolphinDB 等。

由於用 Docker，以上幾種資料庫都很容易架設，網路上可以找到現成的
Docker Image，像本書架設 MySQL，就是使用現成的 Docker Image。筆
者皆架設後，比較 insert 輸入、select 輸出這兩個最常使用的語法，最後
DolphinDB 勝出，主要是效能上的優勢，當然 DolphinDB 也不是完全沒
有缺點，主要還是看使用情境。如果需求是好上手、通用性，那可以選
擇 MySQL，如果需求是效能，那可以選用專門的時序資料庫。

當然實務上，不會只選擇單一資料庫，使用多種資料庫是可行的。以
FinMind 為例，使用 MySQL、Redis、DolphinDB 三種資料庫，依序處理
結構化資料、即時資料、時間序列資料，根據不同情境，進行調整。

有資料庫之後，可以將上一個章節的三隻爬蟲，抓到的資料，寫進資料
庫儲存，在下個章節，會將介紹上傳資料庫的程式架構。

5.4 上傳資料到資料庫

5.4.1 架構介紹

https://github.com/FinMind/FinMindBook/tree/master/DataEngineering/Chapter5/5.4

```
├── create_partition_table.sql
├── create_table.sql
├── financialdata
│   ├── clients.py
│   ├── __init__.py
│   ├── router.py
│   ├── taifex_crawler.py
│   ├── tpex_crawler.py
│   └── twse_crawler.py
├── financialdata.egg-info
│   ├── dependency links.txt
│   ├── PKG-INFO
│   ├── SOURCES.txt
│   └── top_level.txt
├── Pipfile
├── Pipfile.lock
├── README.md
└── setup.py
```

架構與 5.2.1 類似，一樣有 Pipfile，不同的是，現在多了 setup.py，這個檔案主要是讓你的專案變成一個 module，可以直接 import

https://github.com/FinMind/FinMindBook/blob/master/DataEngineering/Chapter5/5.4/setup.py setup.py

```python
from setuptools import setup, find_packages
from os import path
from io import open

here = path.abspath(path.dirname(__file__))

with open(path.join(here, "README.md"), encoding="utf-8") as f:
    long_description = f.read()

setup(
    name="financialdata",  # Required
    version="1.0.1",  # Required
    description="financial mining",  # Optional
    long_description=long_description,  # Optional
    long_description_content_type="text/markdown",
    # Optional (see note above)
    url="https://github.com/linsamtw",  # Optional
    author="linsam",  # Optional
    author_email="samlin266118@gmail.com",  # Optional
    classifiers=[  # Optional
        "Development Status :: 3 - Alpha",
        "Intended Audience :: Developers",
        "Topic :: Software Development :: Build Tools",
        "License :: OSI Approved :: MIT License",
        "Programming Language :: Python :: 3.6",
    ],
    keywords="financial, python",  # Optional
    project_urls={  # Optional
        "documentation": "https://linsamtw.github.io/FinMindDoc/",
        "Source": "https://github.com/linsamtw/FinMind",
    },
)
```

那要如何將專案變成 module，加入 Pipfile 呢？使用以下指令

```
pipenv install -e .
```

Pipfile 如下，注意一點，financialdata = {editable = true,path = "."}，就是
將這個專案加入 module

```
[[source]]
name = "pypi"
url = "https://pypi.org/simple"
verify_ssl = true

[dev-packages]

[packages]
pandas = "*"
requests = "*"
pydantic = "*"
sqlalchemy = "*"
loguru = "*"
financialdata = {editable = true,path = "."}
tqdm = "==4.62.3"

[requires]
python_version = "3.6"
```

除此之外，多了 clients.py、router.py 這兩個檔案，是本章節，要對資料
庫做操作的程式碼，以下將一一介紹。

5.4.2 Clients、Router 資料庫操作

Clients

主要是管理所有對資料庫的連線，將連線有關的程式獨立成一個檔案管理。以目前來說，雖然只有一個對 MySQL 的連線，但未來不只需要 MySQL，可能還會使用 Redis、Time Series DB 等，也可能使用多個 MySQL，以 FinMind 為例，有超過 10 個對資料庫的 client，因此，先獨立出來，對未來的架構設計上，會有很大的幫助

https://github.com/FinMind/FinMindBook/blob/master/DataEngineering/
Chapter5/5.4/financialdata/clients.py

```python
clients.py

from sqlalchemy import create_engine, engine


def get_mysql_financialdata_conn() -> engine.base.Connection:
    """
    user: root
    password: test
    host: localhost
    port: 3306
    database: financialdata
    如果有實體 IP，以上設定可以自行更改
    """
    address = "mysql+pymysql://root:test@localhost:3306/FinancialData"
    engine = create_engine(address)
    connect = engine.connect()
    return connect
```

本書選用 SQLAlchemy 套件，主要是能同時支援 ORM（Object-Relational Mapping）用物件方式操作資料庫，跟 Raw SQL 直接寫純 SQL，這兩種方式都有對應的使用場景，不侷限在某一個部分，對未來架構設計上保持彈性。

Router

主要管理對資料庫的 connect 連線、upload 上傳、alive 確認連線是否活著等操作。一般來說，爬蟲與資料庫操作，這兩部分會分開來設計。

https://github.com/FinMind/FinMindBook/blob/master/DataEngineering/
Chapter5/5.4/financialdata/router.py

```python
router.py

import time
import typing

from loguru import logger
from sqlalchemy import engine


from financialdata import clients


def check_alive(
    connect: engine.base.Connection,
):
    """在每次使用之前，先確認 connect 是否活著"""
    connect.execute("SELECT 1 + 1")
```

```python
def reconnect(
    connect_func: typing.Callable,
) -> engine.base.Connection:
    """如果連線斷掉，重新連線"""
    try:
        connect = connect_func()
    except Exception as e:
        logger.info(
            f"{connect_func.__name__} reconnect error {e}"
        )
    return connect


def check_connect_alive(
    connect: engine.base.Connection,
    connect_func: typing.Callable,
):
    if connect:
        try:
            check_alive(connect)
            return connect
        except Exception as e:
            logger.info(
                f"{connect_func.__name__} connect, error: {e}"
            )
            time.sleep(1)
            connect = reconnect(
                connect_func
            )
            return check_connect_alive(
                connect, connect_func
            )
```

```
    else:
        connect = reconnect(
            connect_func
        )
        return check_connect_alive(
            connect, connect_func
        )


class Router:
    def __init__(self):
        self._mysql_financialdata_conn = (
            clients.get_mysql_financialdata_conn()
        )


    def check_mysql_financialdata_conn_alive(
        self,
    ):
        self._mysql_financialdata_conn = check_connect_alive(
            self._mysql_financialdata_conn,
            clients.get_mysql_financialdata_conn,
        )
        return (
            self._mysql_financialdata_conn
        )


    @property
    def mysql_financialdata_conn(self):
        """
        使用 property，在每次拿取 connect 時，
        都先經過 check alive 檢查 connect 是否活著
        """
```

```
return (
    self.check_mysql_financialdata_conn_alive()
)
```

5.4.3 建立 table

建立資料表，用來存放爬蟲抓到的資料，以台股股價為例，以下是建立 table 的 SQL，或從以下連結下載

https://github.com/FinMind/FinMindBook/blob/master/DataEngineering/Chapter5/5.4/create_table.sql

建立台股股價表

```
create_table.sql
CREATE TABLE `FinancialData`.`TaiwanStockPrice`(
    `StockID` VARCHAR(10) NOT NULL,
    `TradeVolume` BIGINT NOT NULL,
    `Transaction` INT NOT NULL,
    `TradeValue` BIGINT NOT NULL,
    `Open` FLOAT NOT NULL,
    `Max` FLOAT NOT NULL,
    `Min` FLOAT NOT NULL,
    `Close` FLOAT NOT NULL,
    `Change` FLOAT NOT NULL,
    `Date` DATE NOT NULL,
    PRIMARY KEY(`StockID`, `Date`)
)
```

複製以上 SQL 語法，貼到以下執行，這樣就會建立 table 了。

同樣的，也建立台股期貨表

```
CREATE TABLE `FinancialData`.`TaiwanFuturesDaily`(
    `Date` DATE NOT NULL,
    `FuturesID` VARCHAR(10) NOT NULL,
    `ContractDate` VARCHAR(30) NOT NULL,
    `Open` FLOAT NOT NULL,
    `Max` FLOAT NOT NULL,
    `Min` FLOAT NOT NULL,
    `Close` FLOAT NOT NULL,
    `Change` FLOAT NOT NULL,
    `ChangePer` FLOAT NOT NULL,
    `Volume` FLOAT NOT NULL,
    `SettlementPrice` FLOAT NOT NULL,
    `OpenInterest` INT NOT NULL,
    `TradingSession` VARCHAR(11) NOT NULL,
    PRIMARY KEY(`FuturesID`, `Date`)
)
```

這時可以到 http://localhost:8000/，會顯示以下畫面，是剛剛建立的 table

VARCHAR、INT、FLOAT 是資料型態，根據你的資料作調整。設定 PRIMARY KEY 可以保證資料唯一性，也可以增加查詢效率，股價一般都會使用 StockID、Date 作為查詢條件，因此使用以上兩個變數做 PRIMARY KEY，期貨也是同樣邏輯，接下來，將修改程式，從原先存成 csv，改成上傳資料庫。

5.4.4 爬蟲上傳資料庫

相關程式碼如下，可以到以下連結下載

https://github.com/FinMind/FinMindBook/blob/master/DataEngineering/
Chapter5/5.4/financialdata/twse_crawler.py

```
financialdata/twse_crawler.py
import datetime
import sys
import time
import typing


import pandas as pd
```

```python
import requests
from loguru import logger
from pydantic import BaseModel
from tqdm import tqdm

from financialdata.router import Router


def clear_data(
    df: pd.DataFrame,
) -> pd.DataFrame:
    """資料清理，將文字轉成數字"""
    df["Dir"] = (
        df["Dir"]
        .str.split(">")
        .str[1]
        .str.split("<")
        .str[0]
    )
    df["Change"] = (
        df["Dir"] + df["Change"]
    )
    df["Change"] = (
        df["Change"]
        .str.replace(" ", "")
        .str.replace("X", "")
        .astype(float)
    )
    df = df.fillna("")
    df = df.drop(["Dir"], axis=1)
    for col in [
        "TradeVolume",
```

```
            "Transaction",
            "TradeValue",
            "Open",
            "Max",
            "Min",
            "Close",
            "Change",
    ]:
        df[col] = (
            df[col]
            .astype(str)
            .str.replace(",", "")
            .str.replace("X", "")
            .str.replace("+", "")
            .str.replace("----", "0")
            .str.replace("---", "0")
            .str.replace("--", "0")
        )
    return df


def colname_zh2en(
    df: pd.DataFrame,
    colname: typing.List[str],
) -> pd.DataFrame:
    """資料欄位轉換, 英文有助於接下來存入資料庫"""
    taiwan_stock_price = {
        "證券代號": "StockID",
        "證券名稱": "",
        "成交股數": "TradeVolume",
        "成交筆數": "Transaction",
        "成交金額": "TradeValue",
```

```
        "開盤價": "Open",
        "最高價": "Max",
        "最低價": "Min",
        "收盤價": "Close",
        "漲跌(+/-)": "Dir",
        "漲跌價差": "Change",
        "最後揭示買價": "",
        "最後揭示買量": "",
        "最後揭示賣價": "",
        "最後揭示賣量": "",
        "本益比": "",
    }
    df.columns = [
        taiwan_stock_price[col]
        for col in colname
    ]
    df = df.drop([""], axis=1)
    return df



def twse_header():
    """網頁瀏覽時, 所帶的 request header 參數, 模仿瀏覽器發送 request"""
    return {
        "Accept": "application/json, text/javascript, */*; q=0.01",
        "Accept-Encoding": "gzip, deflate",
        "Accept-Language": "zh-TW,zh;q=0.9,en-US;q=0.8,en;q=0.7",
        "Connection": "keep-alive",
        "Host": "www.twse.com.tw",
        "Referer": "https://www.twse.com.tw/zh/page/trading/exchange/
MI_INDEX.html",
        "User-Agent": "Mozilla/5.0 (Windows NT 10.0; Win64;
x64) AppleWebKit/537.36 (KHTML, like Gecko) Chrome/71.0.3578.98
```

```python
Safari/537.36",
        "X-Requested-With": "XMLHttpRequest",
    }



def crawler_twse(
    date: str,
) -> pd.DataFrame:
    """
    證交所網址
    https://www.twse.com.tw/zh/page/trading/exchange/MI_INDEX.html
    """
    # headers 中的 Request url
    url = (
        "https://www.twse.com.tw/exchangeReport/MI_INDEX"
        "?response=json&date={date}&type=ALL"
    )
    url = url.format(
        date=date.replace("-", "")
    )
    # 避免被證交所 ban ip, 在每次爬蟲時, 先 sleep 5 秒
    time.sleep(5)
    # request method
    res = requests.get(
        url, headers=twse_header()
    )
    if (
        res.json()["stat"]
        == "很抱歉，沒有符合條件的資料!"
    ):
        # 如果 date 是周末，會回傳很抱歉，沒有符合條件的資料!
        return pd.DataFrame()
```

```python
    # 2009 年以後的資料，股價在 response 中的 data9
    # 2009 年以後的資料，股價在 response 中的 data8
    # 不同格式，在證交所的資料中，是很常見的，
    # 沒資料的情境也要考慮進去，例如現在週六沒有交易，但在 2007 年週六是有
交易的
    try:
        if "data9" in res.json():
            df = pd.DataFrame(
                res.json()["data9"]
            )
            colname = res.json()[
                "fields9"
            ]
        elif "data8" in res.json():
            df = pd.DataFrame(
                res.json()["data8"]
            )
            colname = res.json()[
                "fields8"
            ]
        elif res.json()["stat"] in [
            "查詢日期小於93年2月11日，請重新查詢!",
            "很抱歉，沒有符合條件的資料!",
        ]:
            return pd.DataFrame()
    except BaseException:
        return pd.DataFrame()

    if len(df) == 0:
        return pd.DataFrame()
    # 欄位中英轉換
    df = colname_zh2en(
```

```python
        df.copy(), colname
    )
    df["Date"] = date
    return df



class TaiwanStockPrice(BaseModel):
    StockID: str
    TradeVolume: int
    Transaction: int
    TradeValue: int
    Open: float
    Max: float
    Min: float
    Close: float
    Change: float
    Date: str



def check_schema(
    df: pd.DataFrame,
) -> pd.DataFrame:
    """檢查資料型態，確保每次要上傳資料庫前，型態正確"""
    df_dict = df.to_dict("records")
    df_schema = [
        TaiwanStockPrice(**dd).__dict__
        for dd in df_dict
    ]
    df = pd.DataFrame(df_schema)
    return df
```

```python
def gen_date_list(
    start_date: str, end_date: str
) -> typing.List[str]:
    """建立時間列表，用於爬取所有資料"""
    start_date = (
        datetime.datetime.strptime(
            start_date, "%Y-%m-%d"
        ).date()
    )
    end_date = (
        datetime.datetime.strptime(
            end_date, "%Y-%m-%d"
        ).date()
    )
    days = (
        end_date - start_date
    ).days + 1
    date_list = [
        str(
            start_date
            + datetime.timedelta(
                days=day
            )
        )
        for day in range(days)
    ]
    return date_list


def main(
    start_date: str, end_date: str
):
    """證交所寫明，※ 本資訊自民國93年2月11日起提供"""
    date_list = gen_date_list(
```

```
        start_date, end_date
    )
    db_router = Router()
    for date in tqdm(date_list):
        logger.info(date)
        df = crawler_twse(date=date)
        if len(df) > 0:
            # 資料清理
            df = clear_data(df.copy())
            # 檢查資料型態
            df = check_schema(df.copy())
            # upload db
            try:
                df.to_sql(
                    name="TaiwanStockPrice",
                    con=db_router.mysql_financialdata_conn,
                    if_exists="append",
                    index=False,
                    chunksize=1000,
                )
            except Exception as e:
                logger.info(e)


if __name__ == "__main__":
    start_date, end_date = sys.argv[1:]
    main(start_date, end_date)
```

主要使用 dataframe 的 to_sql 指令，上傳資料到資料庫，其他部分沒有做
更動。因為在創建 table 時，有設定 PRIMARY KEY，如果同樣的資料已
經存在在 table，會有 error，因此使用 try except 做例外處理。

一樣，先安裝 Package。

```
pipenv sync
```

當到新的章節、不同資料夾時，因為使用的 Package 不同，都請先執行以上指令。

執行以下指令，會進行爬蟲，並將資料存到資料庫中

```
pipenv run python financialdata/twse_crawler.py 2021-01-01 2021-01-10
```

之後切換到 http://localhost:8000/，就會看到爬蟲收集下來的資料了

到這裡，成功使用資料庫，管理爬蟲收集到的資料，未來預計所有爬蟲抓到的資料，都用資料庫管理，相對來說也比 csv 好管理。

本書到此，有證交所、櫃買中心、期交所的爬蟲，爬蟲方式大致上都相同，因此爬蟲不是問題，同時也用資料庫管理抓到的資料，預期未來可能需要收集 20、30 種不同資料，這時需要頻繁去證交所、期交所等做爬蟲，可能會被 ban，本書使用 sleep 5 秒作處理，這時會拖累爬蟲效率，

下一個章節,將使用分散式,處理效率問題。

一次爬蟲要等 5 秒,那分散在 5 台機器上,平均下來,一次只需等待 1 秒,這就是分散的好處。

5.5　分散式爬蟲

5.5.1　為什麼需要分散式?

當你做少量爬蟲時,不太需要分散式架構,單一機器即可,但目標是大量的金融資料爬蟲,因此,單一機器會遇到以下問題。

單一機器負荷不了。基本上,爬蟲程式一多,一定會遇到需要多機器的情況,這時分散式可以有效的高度拓展,同時使用 10 台、20 台機器分擔爬蟲任務。

過度爬蟲被 ban IP。在做證交所、櫃買中心的爬蟲時,大多數人遇到的問題,都是過度爬蟲,被封鎖 IP,這時可能使用 sleep 去解決這問題,但這就影響效率了,因此分散式,可以提升爬蟲的效能。

例如同一時間,用 5 台機器不同 IP 去做爬蟲,這時效率就提升 5 倍,也不用擔心 sleep 的問題。

單一機器故障。如果單一機器突然故障,那至少還有其他分散式機器可以負荷工作,不至於整個爬蟲掛掉。

5.5.2 分散式任務轉發 --- RabbitMQ

這裡介紹 RabbitMQ 這個任務轉發工具，來進行分散式。在分散式爬蟲架構中，會有三個角色。

Producer，主要發派爬蟲任務給 Broker。

Broker，負責從 Producer 接收任務，並轉發任務給 Worker，這裡使用 RabbitMQ 作為訊息傳遞中心。

Worker，工人，負責從 Broker 接收任務，並且執行任務，以本書為例，就是負責做爬蟲的工作，在架構中，會在不同機器開啟多個 Worker 做爬蟲，達到多 IP、任務分散的效果，因此也不用擔心被證交所 ban IP。

Producer →發送任務→ Broker ←拿取任務← Worker

以下，將使用 Docker 一鍵安裝 RabbitMQ，同時並安裝 Flower 這個 Worker 狀態監控工具。

將以下程式，存成 rabbitmq.yml，或是從以下連結下載：

https://github.com/FinMind/FinMindBook/tree/master/DataEngineering/Chapter5/5.5/5.5.2

```
rabbitmq.yml
version: '3'
services:

  rabbitmq:
    image: 'rabbitmq:3.6-management-alpine'
    ports:
      # docker publish port 5672/15672 to 5672/15672
      # 將 docker 內部 ip 5672/15672, 跟外部 5672/15672 做連結
```

```
    - '5672:5672'
    - '15672:15672'
  environment:
    RABBITMQ_DEFAULT_USER: "worker"
    RABBITMQ_DEFAULT_PASS: "worker"
    RABBITMQ_DEFAULT_VHOST: "/"
  networks:
    - dev


flower:
  image: mher/flower:0.9.5
  command: ["flower", "--broker=amqp://worker:worker@rabbitmq",
"--port=5555"]
  ports:
    # docker publish port 5555 to 5555
    # 將 docker 內部 ip 5555, 跟外部 5555 做連結
    - 5555:5555
  depends_on:
    - rabbitmq
  networks:
    - dev

networks:
  dev:
```

接著，一樣的指令啟動 RabbitMQ，

```
docker-compose -f rabbitmq.yml up -d
```

打開瀏覽器，輸入 http://localhost:15672/，帳號密碼是 worker/worker，
就會看到以下畫面，這就是 RabbitMQ 的畫面：

再打開瀏覽器，輸入 http://localhost:5555，就可以看到 Flower 的畫面，
Flower 是負責監控工人 Worker 的狀態，看 Worker 是否活著，工作是否
正常，目前還沒有啟動 Worker，因此看不到東西。

在下個章節，將進入主題，使用爬蟲搭配分散式架構，並且上傳資料到
資料庫。

5.5.3 Python 分散式爬蟲工具 --- Celery

本章節，使用 Celery 這個 Python 的分散式套件，在爬蟲之前，先舉個
Celery 的簡單範例。

整份程式碼，可以從以下連結下載：

https://github.com/FinMind/FinMindBook/tree/master/DataEngineering/
Chapter5/5.5/5.5.3

建立 worker.py，待會會使用這個檔案啟動工人 Worker，Worker 連線
RabbitMQ 之後，只要有任務，工人就會做事。

https://github.com/FinMind/FinMindBook/blob/master/DataEngineering/
Chapter5/5.5/5.5.3/worker.py

```
worker.py

from celery import Celery


app = Celery(
    "task",
    # 只包含 tasks.py 裡面的程式, 才會成功執行
    include=["tasks"],
    # 連線到 rabbitmq,
    # pyamqp://user:password@localhost:5672/
    # 本書設定的帳號密碼都是 worker
    broker="pyamqp://worker:worker@localhost:5672/",
)
```

建立 tasks.py，註冊 crawler 函數，有註冊的函數，Producer 才能發送。

以下程式印出 crawler、upload db 字串，未來將在這兩個地方，結合 5.4
章節，改成爬蟲與上傳資料庫的程式，目前暫時先用 print 代替。

https://github.com/FinMind/FinMindBook/blob/master/DataEngineering/
Chapter5/5.5/5.5.3/tasks.py

tasks.py

```python
from worker import app
# 註冊 task, 有註冊的 task 才可以變成任務發送給 rabbitmq
@app.task()
def crawler(x):
    print("crawler")
    print("upload db")
    return x
```

這裡寫法比較特別，使用 crawler.delay 發送任務，x=0 是這個函數所需的參數。

https://github.com/FinMind/FinMindBook/blob/master/DataEngineering/Chapter5/5.5/5.5.3/producer.py

producer.py

```python
from tasks import crawler

# 發送任務有兩種方式
# 1.
crawler.delay(x=0)
# 2.
# task = crawler.s(x=0)
# task.apply_async()
```

安裝 Package

```
pipenv sync
```

啟動 Celery Worker

```
pipenv run celery -A worker worker --loglevel=info
```

如果執行畫面與下圖相同，代表成功啟動：

```
PROBLEMS    OUTPUT    DEBUG CONSOLE    TERMINAL                                    1: python        ∨   +  ⬚  🗑  ∧

sam@DESKTOP-IKT69L5:~/FinMindBook/DataEngineering/Chapter5/5.5/5.5.3$ pipenv run celery -A worker worker --loglevel=info

 ------------ celery@DESKTOP-IKT69L5 v5.0.5 (singularity)
--- ***** -----
-- ******* ---- Linux-4.4.0-19041-Microsoft-x86_64-with-Ubuntu-18.04-bionic 2021-10-30 17:03:01
- *** --- * ---
- ** ---------- [config]
- ** ---------- .> app:         task:0x7f61279fe4e0
- ** ---------- .> transport:   amqp://worker:**@localhost:5672//
- ** ---------- .> results:     disabled://
- *** --- * --- .> concurrency: 12 (prefork)
-- ******* ---- .> task events: OFF (enable -E to monitor tasks in this worker)
--- ***** -----
 ------------ [queues]
                .> celery           exchange=celery(direct) key=celery


[tasks]
  . tasks.crawler

[2021-10-30 17:03:01,641: INFO/MainProcess] Connected to amqp://worker:**@127.0.0.1:5672//
[2021-10-30 17:03:01,668: INFO/MainProcess] mingle: searching for neighbors
[2021-10-30 17:03:02,713: INFO/MainProcess] mingle: all alone
[2021-10-30 17:03:02,760: INFO/MainProcess] celery@DESKTOP-IKT69L5 ready.
[2021-10-30 17:03:05,830: INFO/MainProcess] Events of group {task} enabled by remote.
```

打開 flower http://localhost:5555/ 畫面，就會看到剛剛的 Worker 的狀態，
Status 是 Online

如果關閉 Celery，Status 就會變成 Offline，這就是 Flower 的用途，用來
監控工人的狀態。

接下來，使用以下指令，發送任務。

```
pipenv run python producer.py
```

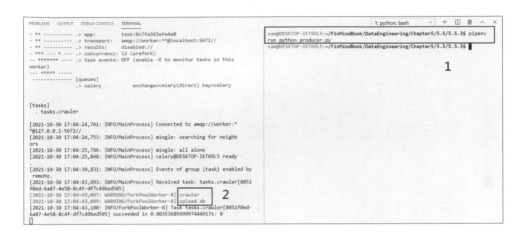

Worker 接到任務之後，就會開始做事，印出 crawler、upload db 這兩個字串，到這邊，已經成功使用 RabbitMQ 分散式架構了。

那 RabbitMQ 用在哪呢？這時先把 Worker 關掉，並且再發一次任務，並打開 http://localhost:15672/

點選 **queue**，底下出現 Celery 的 Name，這就是剛剛發送的任務，再點選 **Celery**，跳轉頁面後，在 **Get messages** 部分，點選 **Get Message(s)** 的按鈕，可以看到這個 **task** 的訊息，特別是 **task** 與 **kwargsrepr** 這兩部分，**task** 顯示任務是要執行 **Tasks.crawler** 這個函數，**kwargsrepr** 中的 {'x': 0}，是這個任務的參數，對比上面的程式碼，應該就很清楚了。

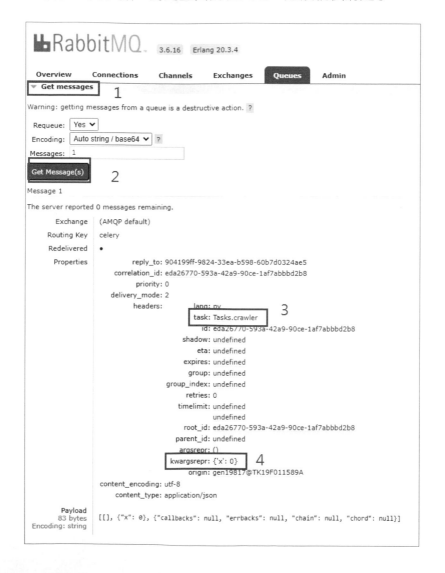

本章節用個簡單例子展示 RabbitMQ、Celery 這個工具，讀者只要將印出 crawler、upload db 的這兩部分，改成爬蟲、上傳資料庫的程式碼，就可以順利達到分散式爬蟲的效果。

5.5.4 真實應用場景

在 5.5.3 中，Worker、RabbitMQ 都是在同一台機器上，真正要達到分散的場景，至少要準備兩台機器，這時一般會遇到固定 IP 和機器不夠的問題，可以參考 Chapter4 中的 Linode 介紹，一個月只要 5 美金，開兩台機器一個月只要 10 美金，約 300 台幣，比單純租網路還便宜，以下示範多 ip 的設定。

請先到 Linode 申請 2 組 5 美金的機器，Ubuntu 系統，這時你會得到 2 組 IP，本書用 rabbitmq_ip、worker1_ip 代替。

設定 Linode 步驟

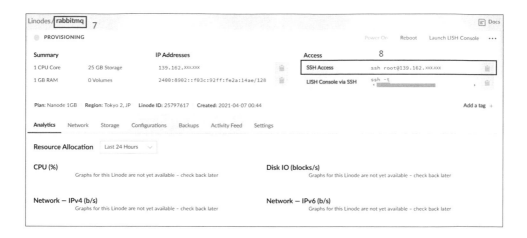

分別 ssh 到以上兩台機器，安裝所需套件、環境設定。

安裝 Docker、pyenv：

```
apt-get update
apt-get install -y docker.io
apt-get install curl git bzip2 -y
curl https://pyenv.run | bash
```

環境設定：

```
echo 'export LC_ALL=C.UTF-8' >> ~/.bashrc
echo 'export LANG=C.UTF-8' >> ~/.bashrc
echo 'export PYENV_ROOT="$HOME/.pyenv"' >> ~/.bashrc
echo 'export PATH="$PYENV_ROOT/shims:$PATH"' >> ~/.bashrc
echo 'export PATH="$PYENV_ROOT/bin:$PATH"' >> ~/.bashrc
echo -e 'if command -v pyenv 1>/dev/null 2>&1; then\n  eval "$(pyenv
init -)"\nfi' >> ~/.bashrc
exec $SHELL
```

安裝 miniconda3-4.3.30：

```
pyenv install miniconda3-4.3.30
pyenv global miniconda3-4.3.30
```

clone 程式碼、安裝套件：

```
pip install docker-compose pipenv
git clone https://github.com/FinMind/FinMindBook.git
cd /FinMindBook/DataEngineering/Chapter5/5.5/5.5.4/
pipenv sync
```

到 rabbitmq_ip 的機器，架設 RabbitMQ：

```
cd /FinMindBook/DataEngineering/Chapter5/5.5/5.5.4/
docker-compose -f rabbitmq.yml up -d
```

這時瀏覽 http://rabbitmq_ip:15672/，會看到以下畫面：

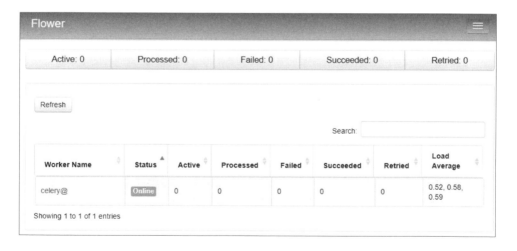

再到 worker1_ip 的機器，啟動 Celery

```
cd /FinMindBook/DataEngineering/Chapter5/5.5/5.5.4/
pipenv run celery -A worker worker --loglevel=info
```

這時到 http://rabbitmq_ip:5555/ 會看到以下畫面

發送任務

```
pipenv run python producer.py
```

成功執行

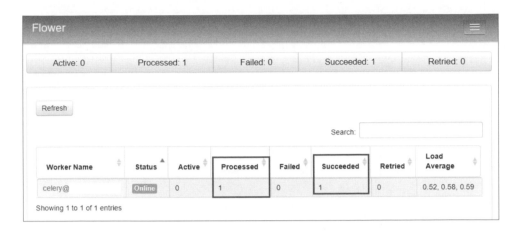

這時就達到分散式了，你可以再多開幾台 Worker，讓 Worker 去處理你的爬蟲任務，並使用 5.3 教學，在雲端架設資料庫，將爬蟲抓到的資料，存在資料庫。

本書到此，已經能成功在雲端上使用分散式爬蟲，並能高度拓展，而資料庫也成功架設了，只要持續新增爬蟲，就能簡單的收集各種資料，下一個章節，將會擷取部分 FinMind 架構做介紹，統整爬蟲、資料庫、分散式。

5.5.5 FinMind 分散式爬蟲架構

可以到以下連結下載程式碼，

https://github.com/FinMind/FinMindBook/tree/master/DataEngineering/Chapter5/5.5/5.5.5

擷取部分 FinMind 爬蟲架構來進行展示。

```
├──── create_partition_table.sql
├──── financialdata
│    ├──── backend
│    │    ├──── db
│    │    │    ├──── clients.py
│    │    │    ├──── db.py
│    │    │    ├──── __init__.py
│    │    │    └──── router.py
│    │    └──── __init__.py
│    ├──── config.py
│    ├──── crawler
│    │    ├──── __init__.py
│    │    ├──── taiwan_futures_daily.py
│    │    └──── taiwan_stock_price.py
│    ├──── __init__.py
│    ├──── producer.py
│    ├──── schema
│    │    ├──── dataset.py
│    │    └──── __init__.py
│    └──── tasks
│         ├──── __init__.py
│         ├──── task.py
│         └──── worker.py
├──── genenv.py
├──── local.ini
├──── Makefile
├──── mysql.yml
├──── Pipfile
├──── Pipfile.lock
├──── rabbitmq.yml
├──── README.md
└──── setup.py
```

mysql.yml、rabbitmq.yml 分別架設相對應的服務，其餘架構主要分成幾個部分，**backend**、**crawler**、**schema**、**tasks**、**config.py**、**producer.py**，

Backend 資料夾裡，主要是 5.4.2 提到過的 clients.py、db.py、router.py，
管理對資料庫的操作：

https://github.com/FinMind/FinMindBook/blob/master/DataEngineering/
Chapter5/5.5/5.5.5/financialdata/backend/db/clients.py

clients.py，建立對資料庫的連線。

```python
from financialdata.config import (
    MYSQL_DATA_USER,
    MYSQL_DATA_PASSWORD,
    MYSQL_DATA_HOST,
    MYSQL_DATA_PORT,
    MYSQL_DATA_DATABASE,
)
from sqlalchemy import create_engine, engine


def get_mysql_financialdata_conn() -> engine.base.Connection:
    address = (
        f"mysql+pymysql://{MYSQL_DATA_USER}:{MYSQL_DATA_PASSWORD}"
        f"@{MYSQL_DATA_HOST}:{MYSQL_DATA_PORT}/{MYSQL_DATA_DATABASE}"
    )
    engine = create_engine(address)
    connect = engine.connect()
    return connect
```

https://github.com/FinMind/FinMindBook/blob/master/DataEngineering/
Chapter5/5.5/5.5.5/financialdata/backend/db/db.py

db.py 對資料庫操作，其中要注意的是，這裡使用兩種方式上傳資料到資
料庫：

1. 用 pandas 原生方式上傳
2. 用 SQL 語法上傳

pandas 的方式上傳速度較快，但遇到重複資料，而 table 又有設主鍵時，會發生錯誤，這時需要寫 SQL 語法才能正確上傳，主要是 SQL 語法中特別加入一段，DUPLICATE KEY UPDATE，可以在遇到重複資料時，改成 update 更新的方式上傳資料，藉此避免因為重複資料，導致資料上傳失敗的問題。

```python
import typing

import pandas as pd
import pymysql
from loguru import logger
from sqlalchemy import engine


def update2mysql_by_pandas(
    df: pd.DataFrame,
    table: str,
    mysql_conn: engine.base.Connection,
):
    if len(df) > 0:
        try:
            df.to_sql(
                name=table,
                con=mysql_conn,
                if_exists="append",
                index=False,
                chunksize=1000,
            )
        except Exception as e:
```

```
            pass


def build_update_sql(
    colname: typing.List[str],
    value: typing.List[str],
):
    update_sql = ",".join(
        [
            ' `{}` = "{}" '.format(
                colname[i],
                str(value[i]),
            )
            for i in range(len(colname))
            if str(value[i])
        ]
    )
    return update_sql


def build_df_update_sql(
    table: str, df: pd.DataFrame
) -> typing.List[str]:
    logger.info("build_df_update_sql")
    df_columns = list(df.columns)
    sql_list = []
    for i in range(len(df)):
        temp = list(df.iloc[i])
        value = [
            pymysql.converters.escape_string(
                str(v)
            )
```

```
            for v in temp
        ]
        sub_df_columns = [
            df_columns[j]
            for j in range(len(temp))
        ]
        update_sql = build_update_sql(
            sub_df_columns, value
        )
        # SQL 上傳資料方式
        # DUPLICATE KEY UPDATE 意思是
        # 如果有重複，就改用 update 的方式
        # 避免重複上傳
        sql = """INSERT INTO `{}`({})VALUES ({}) ON DUPLICATE KEY UPDATE {}
            """.format(
            table,
            "`{}`".format(
                "`,`".join(
                    sub_df_columns
                )
            ),
            '"{}"'.format(
                '","'.join(value)
            ),
            update_sql,
        )
        sql_list.append(sql)
    return sql_list


def update2mysql_by_sql(
    df: pd.DataFrame,
```

```python
    table: str,
    mysql_conn: engine.base.Connection,
):
    sql = build_df_update_sql(table, df)
    commit(
        sql=sql, mysql_conn=mysql_conn
    )


def commit(
    sql: typing.Union[
        str, typing.List[str]
    ],
    mysql_conn: engine.base.Connection = None,
):
    logger.info("commit")
    try:
        trans = mysql_conn.begin()
        if isinstance(sql, list):
            for s in sql:
                try:
                    mysql_conn.execution_options(
                        autocommit=False
                    ).execute(
                        s
                    )
                except Exception as e:
                    logger.info(e)
                    logger.info(s)
                    break

        elif isinstance(sql, str):
```

```
            mysql_conn.execution_options(
                autocommit=False
            ).execute(
                sql
            )
        trans.commit()
    except Exception as e:
        trans.rollback()
        logger.info(e)


def upload_data(
    df: pd.DataFrame,
    table: str,
    mysql_conn: engine.base.Connection,
):
    if len(df) > 0:
        # 直接上傳
        if update2mysql_by_pandas(
            df=df,
            table=table,
            mysql_conn=mysql_conn,
        ):
            pass
        else:
            # 如果有重複的資料
            # 使用 SQL 語法上傳資料
            update2mysql_by_sql(
                df=df,
                table=table,
                mysql_conn=mysql_conn,
            )
```

https://github.com/FinMind/FinMindBook/blob/master/DataEngineering/
Chapter5/5.5/5.5.5/financialdata/backend/db/router.py

router.py 管理 client，並檢查 connect 連線是否活著。

```python
import time
import typing

from loguru import logger
from sqlalchemy import engine
from financialdata.backend.db import clients


def check_alive(connect: engine.base.Connection):
    connect.execute("SELECT 1 + 1")


def check_connect_alive(
    connect: engine.base.Connection,
    connect_func: typing.Callable,
):
    if connect:
        try:
            check_alive(connect)
            return connect
        except Exception as e:
            logger.info(
                f"""
                {connect_func.__name__} reconnect, error: {e}
                """
            )
            time.sleep(1)
```

```
        try:
            connect = connect_func()
        except Exception as e:
            logger.info(
                f"""
                {connect_func.__name__} connect error, error: {e}
                """
            )
        return check_connect_alive(connect, connect_func)


class Router:
    def __init__(self):
        self._mysql_financialdata_conn = clients.get_mysql_financialdata_
conn()

    def check_mysql_financialdata_conn_alive(self):
        self._mysql_financialdata_conn = check_connect_alive(
            self._mysql_financialdata_conn,
            clients.get_mysql_financialdata_conn,
        )
        return self._mysql_financialdata_conn

    @property
    def mysql_financialdata_conn(self):
        return self.check_mysql_financialdata_conn_alive()

    def close_connection(self):
        self._mysql_financialdata_conn.close()
```

crawler 資料夾中，是 taiwan_stock_price.py 和 taiwan_futures_daily.py，
台股股價爬蟲、台股期貨爬蟲，大致上與 5.4 程式的相同，主要是合併證

交所爬蟲 twse_crawler.py 與櫃買中心爬蟲 tpex_crawler.py，統整為台股
股價爬蟲 taiwan_stock_price.py，程式如下：

https://github.com/FinMind/FinMindBook/blob/master/DataEngineering/
Chapter5/5.5/5.5.5/financialdata/crawler/taiwan_stock_price.py

```
taiwan_stock_price.py
import datetime
import time
import typing

import pandas as pd
import requests
from loguru import logger
from financialdata.schema.dataset import (
    check_schema,
)


def is_weekend(day: int) -> bool:
    return day in [0, 6]


def gen_task_paramter_list(
    start_date: str, end_date: str
) -> typing.List[str]:
    start_date = (
        datetime.datetime.strptime(
            start_date, "%Y-%m-%d"
        ).date()
    )
    end_date = (
```

```python
        datetime.datetime.strptime(
            end_date, "%Y-%m-%d"
        ).date()
    )
    days = (
        end_date - start_date
    ).days + 1
    date_list = [
        start_date
        + datetime.timedelta(days=day)
        for day in range(days)
    ]
    # 排除掉周末非交易日
    date_list = [
        dict(
            date=str(d),
            data_source=data_source,
        )
        for d in date_list
        for data_source in [
            "twse",
            "tpex",
        ]
        if not is_weekend(d.weekday())
    ]
    return date_list


def clear_data(
    df: pd.DataFrame,
) -> pd.DataFrame:
    """資料清理，將文字轉成數字"""
```

```python
    for col in [
        "TradeVolume",
        "Transaction",
        "TradeValue",
        "Open",
        "Max",
        "Min",
        "Close",
        "Change",
    ]:
        df[col] = (
            df[col]
            .astype(str)
            .str.replace(",", "")
            .str.replace("X", "")
            .str.replace("+", "")
            .str.replace("----", "0")
            .str.replace("---", "0")
            .str.replace("--", "0")
            .str.replace(" ", "")
            .str.replace("除權息", "0")
            .str.replace("除息", "0")
            .str.replace("除權", "0")
        )
    return df



def colname_zh2en(
    df: pd.DataFrame,
    colname: typing.List[str],
) -> pd.DataFrame:
    """資料欄位轉換，英文有助於接下來存入資料庫"""
```

```python
    taiwan_stock_price = {
        "證券代號": "StockID",
        "證券名稱": "",
        "成交股數": "TradeVolume",
        "成交筆數": "Transaction",
        "成交金額": "TradeValue",
        "開盤價": "Open",
        "最高價": "Max",
        "最低價": "Min",
        "收盤價": "Close",
        "漲跌(+/-)": "Dir",
        "漲跌價差": "Change",
        "最後揭示買價": "",
        "最後揭示買量": "",
        "最後揭示賣價": "",
        "最後揭示賣量": "",
        "本益比": "",
    }
    df.columns = [
        taiwan_stock_price[col]
        for col in colname
    ]
    df = df.drop([""], axis=1)
    return df


def twse_header():
    """網頁瀏覽時, 所帶的 request header 參數, 模仿瀏覽器發送 request"""
    return {
        "Accept": "application/json, text/javascript, */*; q=0.01",
        "Accept-Encoding": "gzip, deflate",
        "Accept-Language": "zh-TW,zh;q=0.9,en-US;q=0.8,en;q=0.7",
```

```
        "Connection": "keep-alive",
        "Host": "www.twse.com.tw",
        "Referer": "https://www.twse.com.tw/zh/page/trading/exchange/
MI_INDEX.html",
        "User-Agent": "Mozilla/5.0 (Windows NT 10.0; Win64;
x64) AppleWebKit/537.36 (KHTML, like Gecko) Chrome/71.0.3578.98
Safari/537.36",
        "X-Requested-With": "XMLHttpRequest",
    }


def tpex_header():
    """網頁瀏覽時，所帶的 request header 參數，模仿瀏覽器發送 request"""
    return {
        "Accept": "application/json, text/javascript, */*; q=0.01",
        "Accept-Encoding": "gzip, deflate",
        "Accept-Language": "zh-TW,zh;q=0.9,en-US;q=0.8,en;q=0.7",
        "Connection": "keep-alive",
        "Host": "www.tpex.org.tw",
        "Referer": "https://www.tpex.org.tw/web/stock/aftertrading/
otc_quotes_no1430/stk_wn1430.php?l=zh-tw",
        "User-Agent": "Mozilla/5.0 (Windows NT 10.0; Win64; x64)
AppleWebKit/537.36 (KHTML, like Gecko) Chrome/73.0.3683.103
Safari/537.36",
        "X-Requested-With": "XMLHttpRequest",
    }


def set_column(
    df: pd.DataFrame,
) -> pd.DataFrame:
    """設定資料欄位名稱"""
```

```python
    df.columns = [
        "StockID",
        "Close",
        "Change",
        "Open",
        "Max",
        "Min",
        "TradeVolume",
        "TradeValue",
        "Transaction",
    ]
    return df


def crawler_tpex(
    date: str,
) -> pd.DataFrame:
    """
    櫃買中心網址
    https://www.tpex.org.tw/web/stock/aftertrading/otc_quotes_no1430/
stk_wn1430.php?l=zh-tw
    """
    logger.info("crawler_tpex")
    # headers 中的 Request url
    url = "https://www.tpex.org.tw/web/stock/aftertrading/otc_quotes_
no1430/stk_wn1430_result.php?l=zh-tw&d={date}&se=AL"
    url = url.format(
        date=convert_date(date)
    )
    # 避免被櫃買中心 ban ip, 在每次爬蟲時, 先 sleep 5 秒
    time.sleep(5)
    # request method
```

```
    res = requests.get(
        url, headers=tpex_header()
    )
    data = res.json().get("aaData", [])
    df = pd.DataFrame(data)
    if not data or len(df) == 0:
        return pd.DataFrame()
    # 櫃買中心回傳的資料，並無資料欄位，因此這裡直接用 index 取特定欄位
    df = df[[0, 2, 3, 4, 5, 6, 7, 8, 9]]
    # 欄位中英轉換
    df = set_column(df.copy())
    df["Date"] = date
    df = clear_data(df.copy())
    return df


def crawler_twse(
    date: str,
) -> pd.DataFrame:
    """
    證交所網址
    https://www.twse.com.tw/zh/page/trading/exchange/MI_INDEX.html
    """
    logger.info("crawler_twse")
    # headers 中的 Request url
    url = "https://www.twse.com.tw/exchangeReport/MI_INDEX?response=
json&date={date}&type=ALL"
    url = url.format(
        date=date.replace("-", "")
    )
    # 避免被證交所 ban ip, 在每次爬蟲時, 先 sleep 5 秒
    time.sleep(5)
```

```python
# request method
res = requests.get(
    url, headers=twse_header()
)
# 2009 年以後的資料, 股價在 response 中的 data9
# 2009 年以後的資料, 股價在 response 中的 data8
# 不同格式, 在證交所的資料中, 是很常見的,
# 沒資料的情境也要考慮進去, 例如現在週六沒有交易,但在 2007 年週六是有
交易的
df = pd.DataFrame()
try:
    if "data9" in res.json():
        df = pd.DataFrame(
            res.json()["data9"]
        )
        colname = res.json()[
            "fields9"
        ]
    elif "data8" in res.json():
        df = pd.DataFrame(
            res.json()["data8"]
        )
        colname = res.json()[
            "fields8"
        ]
    elif res.json()["stat"] in [
        "查詢日期小於93年2月11日,請重新查詢!",
        "很抱歉,沒有符合條件的資料!",
    ]:
        pass
except Exception as e:
    logger.error(e)
```

```python
        return pd.DataFrame()

    if len(df) == 0:
        return pd.DataFrame()
    # 欄位中英轉換
    df = colname_zh2en(
        df.copy(), colname
    )
    df["Date"] = date
    df = convert_change(df.copy())
    df = clear_data(df.copy())
    return df


def convert_change(
    df: pd.DataFrame,
) -> pd.DataFrame:
    logger.info("convert_change")
    df["Dir"] = (
        df["Dir"]
        .str.split(">")
        .str[1]
        .str.split("<")
        .str[0]
    )
    df["Change"] = (
        df["Dir"] + df["Change"]
    )
    df["Change"] = (
        df["Change"]
        .str.replace(" ", "")
        .str.replace("X", "")
```

```python
            .astype(float)
        )
    df = df.fillna("")
    df = df.drop(["Dir"], axis=1)
    return df



def convert_date(date: str) -> str:
    logger.info("convert_date")
    year, month, day = date.split("-")
    year = int(year) - 1911
    return f"{year}/{month}/{day}"



def crawler(
    parameter: typing.Dict[
        str,
        typing.List[
            typing.Union[
                str, int, float
            ]
        ],
    ]
) -> pd.DataFrame:
    logger.info(parameter)
    date = parameter.get("date", "")
    data_source = parameter.get(
        "data_source", ""
    )
    if data_source == "twse":
        df = crawler_twse(date)
    elif data_source == "tpex":
```

```
        df = crawler_tpex(date)
    df = check_schema(
        df.copy(),
        dataset="TaiwanStockPrice",
    )
    return df
```

期貨爬蟲只是單純將 taifex_crawler.py 改名為 taiwan_futures_daily.py，程式碼與 5.4 相同，在這就不多做介紹。

schema 資料夾中，是統整所有爬蟲的 schema，統整在一起比較好管理

```
dataset.py
from pydantic import BaseModel
import importlib


import pandas as pd



class TaiwanStockPrice(BaseModel):
    StockID: str
    TradeVolume: int
    Transaction: int
    TradeValue: int
    Open: float
    Max: float
    Min: float
    Close: float
    Change: float
    Date: str
```

```
class TaiwanFuturesDaily(BaseModel):
    Date: str
    FuturesID: str
    ContractDate: str
    Open: float
    Max: float
    Min: float
    Close: float
    Change: float
    ChangePer: float
    Volume: float
    SettlementPrice: float
    OpenInterest: int
    TradingSession: str
```

```
def check_schema(df: pd.DataFrame, dataset: str) -> pd.DataFrame:
    """檢查資料型態, 確保每次要上傳資料庫前, 型態正確"""
    df_dict = df.to_dict("records")
    schema = getattr(
        importlib.import_module("financialdata.schema.dataset"),
        dataset,
    )
    df_schema = [schema(**dd).__dict__ for dd in df_dict]
    df = pd.DataFrame(df_schema)
    return df
```

tasks 資料夾中,是整理 5.5.3 的分散式架構,包含 worker.py、task.py

```
worker.py
```
```
from celery import Celery
from financialdata.config import (
```

```
    WORKER_ACCOUNT,
    WORKER_PASSWORD,
    MESSAGE_QUEUE_HOST,
    MESSAGE_QUEUE_PORT,
)

broker = (
    f"pyamqp://{WORKER_ACCOUNT}:{WORKER_PASSWORD}@"
    f"{MESSAGE_QUEUE_HOST}:{MESSAGE_QUEUE_PORT}/"
)
app = Celery(
    "task",
    include=["financialdata.tasks.task"],
    broker=broker,
)
```

task.py

```
import importlib
import typing

from financialdata.backend import db
from financialdata.tasks.worker import app


# 註冊 task, 有註冊的 task 才可以變成任務發送給 rabbitmq
@app.task()
def crawler(dataset: str, parameter: typing.Dict[str, str]):
    # 使用 getattr, importlib,
    # 根據不同 dataset, 使用相對應的 crawler 收集資料
    # 爬蟲
    df = getattr(
        importlib.import_module(f"financialdata.crawler.{dataset}"),
```

```
    "crawler",
)(parameter=parameter)
# 上傳資料庫
db.upload_data(df, dataset, db.router.mysql_financialdata_conn)
```

producer.py 跟 5.5.3 一樣，是負責發送任務的，這裡預期可以針對不同爬蟲，統一一個規格去發送任務，並將爬蟲任務根據不同日期，切成一個個獨立的爬蟲，才方便使用分散式去獨力完成，因此使用以下設計。

producer.py
```
import importlib
import sys

from loguru import logger

from financialdata.backend import db
from financialdata.tasks.task import crawler


def Update(dataset: str, start_date: str, end_date: str):
    # 拿取每個爬蟲任務的參數列表，
    # 包含爬蟲資料的日期 date，例如 2021-04-10 的台股股價，
    # 資料來源 data_source，例如 twse 證交所、tpex 櫃買中心
    parameter_list = getattr(
        importlib.import_module(f"financialdata.crawler.{dataset}"),
        "gen_task_paramter_list",
    )(start_date=start_date, end_date=end_date)
    # 用 for loop 發送任務
    for parameter in parameter_list:
        logger.info(f"{dataset}, {parameter}")
        task = crawler.s(dataset, parameter)
```

```
        # queue 參數，可以指定要發送到特定 queue 列隊中
        task.apply_async(queue=parameter.get("data_source", ""))


    db.router.close_connection()



if __name__ == "__main__":
    dataset, start_date, end_date = sys.argv[1:]
    Update(dataset, start_date, end_date)
```

統一使用 **gen_task_paramter_list** 產生每個任務的參數列表，並統一使用 crawler 做爬蟲回傳 dataframe，因此規範每隻爬蟲必須要有 **gen_task_paramter_list**、**crawler** 這兩個 function

config.py 是整份程式用到的設定檔，包含對資料庫、**RabbitMQ** 的設定，那為什麼要特別抽出來呢？

```
config.py
import os

MYSQL_DATA_HOST = os.environ.get("MYSQL_DATA_HOST", "127.0.0.1")
MYSQL_DATA_USER = os.environ.get("MYSQL_DATA_USER", "root")
MYSQL_DATA_PASSWORD = os.environ.get("MYSQL_DATA_PASSWORD", "test")
MYSQL_DATA_PORT = int(os.environ.get("MYSQL_DATA_PORT", "3306"))
MYSQL_DATA_DATABASE = os.environ.get("MYSQL_DATA_DATABASE", "FinancialData")

WORKER_ACCOUNT = os.environ.get("WORKER_ACCOUNT", "worker")
WORKER_PASSWORD = os.environ.get("WORKER_PASSWORD", "worker")

MESSAGE_QUEUE_HOST = os.environ.get("MESSAGE_QUEUE_HOST", "127.0.0.1")
MESSAGE_QUEUE_PORT = int(os.environ.get("MESSAGE_QUEUE_PORT", "5672"))
```

第一，是為了方便管理。

第二，在開發與實際產品部屬上，會使用不同設定，開發 dev 就用 localhost，實際部屬 release 就用雲端 IP，甚至在上正式產品環境時，還會經過一層 staging 測試機的環境，因此把環境變數抽出來，有助於未來的開發上，

那環境變數，一般怎麼設定呢？

這裡 FinMind 使用 local.ini 管理開發、測試、正式環境變數：

```ini
local.ini
# dev 環境
[DEFAULT]
MYSQL_DATA_HOST = 127.0.0.1
MYSQL_DATA_USER = root
MYSQL_DATA_PASSWORD = test
MYSQL_DATA_PORT = 3306
MYSQL_DATA_DATABASE = FinancialData
WORKER_ACCOUNT = worker
WORKER_PASSWORD = worker
MESSAGE_QUEUE_HOST = 127.0.0.1
MESSAGE_QUEUE_PORT = 5672


# 測試站環境，這邊可以換成自己測試機的 IP，如果沒有測試機，可以先略過這段
[STAGING]
MYSQL_DATA_HOST = 127.0.0.1
MYSQL_DATA_USER = root
MYSQL_DATA_PASSWORD = test
MYSQL_DATA_PORT = 3306
MYSQL_DATA_DATABASE = FinancialData
WORKER_ACCOUNT = worker
```

```
WORKER_PASSWORD = worker
MESSAGE_QUEUE_HOST = 127.0.0.1
MESSAGE_QUEUE_PORT = 5672


# 正式站環境，這邊可以換成 linode 上的 IP
[RELEASE]
MYSQL_DATA_HOST = 127.0.0.1
MYSQL_DATA_USER = root
MYSQL_DATA_PASSWORD = test
MYSQL_DATA_PORT = 3306
MYSQL_DATA_DATABASE = FinancialData
WORKER_ACCOUNT = worker
WORKER_PASSWORD = worker
MESSAGE_QUEUE_HOST = 127.0.0.1
MESSAGE_QUEUE_PORT = 5672
```

使用 genenv.py 建立 .env 環境檔，搭配 pipenv 使用，pipenv run python
會同時使用 .env 做為環境變數，並且執行 Python，這就不需要額外設定
機器的環境變數：

genenv.py
```
import socket
import os
from configparser import ConfigParser


HOME_PATH = "/".join(os.path.abspath(__file__).split("/")[:-1])
HOST_NAME = socket.gethostname()


local_config = ConfigParser()
local_config.read("local.ini")
if os.environ.get("VERSION", ""):
    section = local_config[os.environ.get("VERSION", "")]
```

```python
elif HOST_NAME in local_config:
    section = local_config[HOST_NAME]
else:
    section = local_config["DEFAULT"]

env_content = ""
for sec in section:
    env_content += "{}={}\n".format(sec.upper(), section[sec])

with open(".env", "w", encoding="utf8") as env:
    env.write(env_content)
```

而新增的 Makefile 是協助紀錄一些常用到的指令，包含啟動資料庫、RabbitMQ、Celery、發送任務、建立環境變數等。

Makefile

```makefile
# 啟動 mysql
create-mysql:
    docker-compose -f mysql.yml up -d


# 啟動 rabbitmq
create-rabbitmq:
    docker-compose -f rabbitmq.yml up -d


# 安裝環境
install-python-env:
    pipenv sync


# 啟動 celery, 專門執行 twse queue 列隊的任務,
run-celery-twse:
    pipenv run celery -A financialdata.tasks.worker worker
--loglevel=info --concurrency=1  --hostname=%h -Q twse
```

```
# 啟動 celery, 專門執行 tpex queue 列隊的任務
run-celery-tpex:
    pipenv run celery -A financialdata.tasks.worker worker
--loglevel=info --concurrency=1  --hostname=%h -Q tpex

# sent task
sent-taiwan-stock-price-task:
    pipenv run python financialdata/producer.py taiwan_stock_price 2021-
04-01 2021-04-12

# 建立 dev 環境變數
gen-dev-env-variable:
    python genenv.py

# 建立 staging 環境變數
gen-staging-env-variable:
    VERSION=STAGING python genenv.py

# 建立 release 環境變數
gen-release-env-variable:
    VERSION=RELEASE python genenv.py
```

安裝 make

```
sudo apt install make
```

使用 make 建立環境變數

```
make gen-dev-env-variable
```

這時就會產生 .env 的檔案，也就是上面提到的。

只需要簡單一個指令,就能任意切換開發、測試、正式環境,是不是很簡單呢?這就是將 config 抽出來的好處:

```
.env
MYSQL_DATA_HOST=127.0.0.1
MYSQL_DATA_USER=root
MYSQL_DATA_PASSWORD=test
MYSQL_DATA_PORT=3306
MYSQL_DATA_DATABASE=FinancialData
WORKER_ACCOUNT=worker
WORKER_PASSWORD=worker
MESSAGE_QUEUE_HOST=127.0.0.1
MESSAGE_QUEUE_PORT=5672
```

使用 make 執行指令,啟動資料庫、RabbitMQ

```
make create-mysql
make create-rabbitmq
```

安裝 Package

```
make install-python-env
```

發送任務

```
make sent-taiwan-stock-price-task
```

啟動 Celery

```
make run-celery-twse
```

這裡 Celery 多出一些額外參數：

concurrency

根據官方文件

> *The number of worker processes/threads can be changed using the --concurrency argument and defaults to the number of CPUs available on the machine.*

意思是同時執行任務的線程數量預設為 CPU 核心數，由於證交所、櫃買中心會 ban IP，因此這裡需額外設定，同時只能做一個任務。

hostname

Worker 的名字，%h 代表的是 hostname。

-Q

代表這個 Celery 會去執行哪些任務列隊，例如 twse/tpex，分別代表這個 Celery 只會執行 twse/tpex 列隊的任務。

而為什麼這裡需要分兩個 Celery？因為實際上，一台機器可以同時執行證交所、櫃買中心爬蟲，因為這是兩個不同網站，同時爬蟲不用擔心太頻繁被 ban。

基本上只要啟動 Celery 後，再發送任務，就會順利將資料爬進 DB 囉。

5.6 定時爬蟲

5.6.1 為什麼需要定時爬蟲？

目前架構上，都是在 command line 上手動發送任務，這樣如果需要每天最新資料，不就需要每天手動執行 command line 了嗎？這非常沒效率，可能部分讀者會使用 crontab，但 crontab 不好做進版控裡，本書希望一切都使用版本控制，以下將介紹 Python 排程管理工具 APScheduler。

5.6.2 APScheduler

APScheduler 是一個 Python 的工作排程工具，由於是 Python 程式碼，可以做進版控，也不限於 Windows 或 Linux，你不用擔心 Windows 環境，沒有 crontab，以下提供一個簡單範例，也可從以下連結下載

https://github.com/FinMind/FinMindBook/blob/master/DataEngineering/Chapter5/5.6/5.6.2/scheduler.py

```
scheduler.py

import time


from apscheduler.schedulers.background import BackgroundScheduler
```

```
from loguru import logger


def sent_crawler_task(dataset: str):
    # 將此段,改成發送任務的程式碼
    logger.info(f"sent_crawler_task {dataset}")


def main():
    scheduler = BackgroundScheduler(timezone="Asia/Taipei")
    # 與crontab類似,設定何時執行,有小時、分鐘、秒參數,* 星號代表任意時間點
    scheduler.add_job(
        id="sent_crawler_task",
        func=sent_crawler_task,
        trigger="cron",
        hour="*",
        minute="*",
        day_of_week="*",
        second="*/5",
        args=["taiwan_stock_price"],
    )
    logger.info("sent_crawler_task")
    scheduler.start()


if __name__ == "__main__":
    main()
    while True:
        time.sleep(600)
```

基本上只要將以上 *logger.info(f"sent_crawler_task {dataset}")* 部分,改成上一章節 update 的程式碼,就能每天定時爬蟲了。

執行方法

```
pipenv run python scheduler.py
```

接下來將 APScheduler 結合 5.5.5 的爬蟲程式。

5.6.3 APScheduelr & 分散式爬蟲架構

APScheduler 結合爬蟲非常簡單，如下：

https://github.com/FinMind/FinMindBook/blob/master/DataEngineering/
Chapter5/5.6/5.6.3/financialdata/scheduler.py

```
scheduler.py
import time
import datetime

from apscheduler.schedulers.background import (
    BackgroundScheduler,
)
from financialdata.producer import (
    Update,
)
from loguru import logger


def sent_crawler_task():
    # 將此段，改成發送任務的程式碼
    # logger.info(f"sent_crawler_task {dataset}")
    today = (
        datetime.datetime.today()
        .date()
```

```python
        .strftime("%Y-%m-%d")
    )
    Update(
        dataset="taiwan_stock_price",
        start_date=today,
        end_date=today,
    )


def main():
    scheduler = BackgroundScheduler(
        timezone="Asia/Taipei"
    )
    # 與crontab類似，設定何時執行，有小時、分鐘、秒參數，* 星號代表任意時間點
    scheduler.add_job(
        id="sent_crawler_task",
        func=sent_crawler_task,
        trigger="cron",
        hour="15",
        minute="0",
        day_of_week="mon-fri",
    )
    logger.info("sent_crawler_task")
    scheduler.start()


if __name__ == "__main__":
    main()
    while True:
        time.sleep(600)
```

這裡 Update 的 start_date、end_date 都設今天,就可以每天更新了,之後只需要手動做一次更新歷史資料即可,如下。

```
pipenv run python financialdata/producer.py taiwan_stock_price 2000-01-
01 2021-06-01
```

另外,scheduler 設定每周一到五,15 點執行,這時基本上證交所、櫃買中心資料都已經更新,如果不放心,可以設 15-17,這樣 15-17 每個整點都會發送一次爬蟲任務,因為 MySQL 有設定 PRIMARY KEY,因此也不用擔心資料重複 insert 問題。

以上是台股股價範例,你可以另外再加上期貨等等的其他爬蟲,基本上架構是一樣的。

本書到此,使用雲端,分散式進行爬蟲,且定時爬蟲。現在假設已經將所有歷史資料,都存入資料庫,那要如何取出呢?如果只有自己存取,寫幾段 SQL 可能就解決了,要給其他人使用呢?基本上,不可能讓對方直接對資料庫做連線,這有資安問題,API 是個不錯的選擇,只要架設好,就可以輕易的給其他人使用,且不限語言,以下將介紹 API 的設計。

資料提供─
RESTful API 設計

6.1 什麼是 API？

API 是一種界接介面，用於不同程式、軟體之間的呼叫、傳遞，自由度高，你可以在 API 做一些限制，對方必須遵守你的規則，才能使用。

而 RESTful API 是一種設計規範，定義了 GET、POST、PUT、DELETE 等操作，例如拿取資料用 GET，推送資料用 POST，更新資料用 PUT，刪除資料用 DELETE，以下將一步步實作 RESTful API。

6.2 輕量 API --- Flask

6.2.1 建立第一個 API

以下是 Flask 官方介紹

Flask is a lightweight WSGI web application framework

簡單來說，Flask 是一個輕量的 Web 框架，可以用於網站架設或是 API 架設，首先，使用 Flask 建立第一個 API

安裝 package

```
pipenv install flask==1.0.0 PyJWT==1.7.1
```

https://github.com/FinMind/FinMindBook/blob/master/DataEngineering/
Chapter6/6.2/app.py

```
app.py
from flask import Flask, jsonify


app = Flask(__name__)



@app.route("/", methods=["GET"])
def hello():
    return jsonify({"Hello": "World"})



if __name__ == "__main__":
    app.run(host="0.0.0.0", port=8888)
```

一樣先安裝 Package

```
pipenv sync
```

執行以下指令

```
pipenv run python app.py
```

會顯示以下畫面

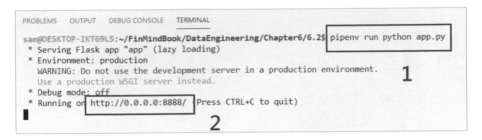

接著到 http://127.0.0.1:8888/，會看到 {"Hello": "World"}，代表成功架設了。

這時再開一個 Python，執行以下指令：

```
import requests

res = requests.get("http://127.0.0.1:8888/")
res.json()
```

就會出現 {"Hello": "World"}

代表 API 架設成功，其他使用者，可以發 request 到以上架設的 API。

Flask 主打的是輕量 Web 框架，並非專門用於 API 架設，那麼接下來，介紹專門用於高效能的 API，Fastapi，它擁有完整的 API 架構，包含良好的架構設計、WebSocket、測試等。

6.3　高效能 API --- FastAPI

6.3.1　建立第一個 API

同樣，引用 FastAPI 官方介紹：

FastAPI is a modern, fast (high-performance), web framework for building APIs with Python 3.6+ based on standard Python type hints.

FastAPI 是一個高效能的 API 框架，快速、簡單，先來一個簡單的例子，安裝 Pacakge

```
pipenv install fastapi==0.63.0 uvicorn==0.14.0
```

程式碼可以從以下連結下載：

https://github.com/FinMind/FinMindBook/blob/master/DataEngineering/Chapter6/6.3/6.3.1/main.py

```
main.py
```
```python
from fastapi import FastAPI


app = FastAPI()



@app.get("/")
def read_root():
    return {"Hello": "World"}



@app.get("/taiwan_stock_price")
def taiwan_stock_price():
    print("get data from mysql")
    return {"data": 123}
```

執行以下指令

```
pipenv run uvicorn main:app --reload --port 8888
```

成功的話，會顯示以下畫面

接著到 http://127.0.0.1:8888/，就會顯示 {"Hello":"World"}。這時模擬使用者，要對這個 API 發送 request，如果畫面與下圖相同，代表成功模擬了，之後只要把 get data from mysql 的部分，改成實際的程式碼，就能完整的串接 API，拿到資料了，下個章節將做個示範。

```
PROBLEMS    OUTPUT    DEBUG CONSOLE    TERMINAL

sam@DESKTOP-IKT69L5:~/FinMindBook/DataEngineering/Chapter6/6.3/6.3.2$ python
Python 3.6.3 |Anaconda, Inc.| (default, Oct 13 2017, 12:02:49)
[GCC 7.2.0] on linux
Type "help", "copyright", "credits" or "license" for more information.
>>> import requests
>>> res = requests.get("http://127.0.0.1:8888/taiwan_stock_price")
>>> res.json()
{'data': 123}
>>>
```

6.3.2 建立 API 回傳真實資料

首先，使用 docker ps 檢查，確保前面爬蟲用到的 MySQL Docker 活著，

```
sam@DESKTOP-IKT69L5:~/FinMindBook/DataEngineering/Chapter6/6.3/6.3.2$ docker ps
CONTAINER ID   IMAGE                             COMMAND                  CREATED       STATUS        PORTS
                                                                                                     NAMES
6ab83448cc06   phpmyadmin/phpmyadmin:5.1.0       "/docker-entrypoint...."  6 hours ago   Up 6 hours    0.0.0.0:8000->80/tcp, :::8000-
>80/tcp                                                                                               53_phpmyadmin_1
bc9a660d7186   mysql:8.0                         "docker-entrypoint.s…"   6 hours ago   Up 6 hours    0.0.0.0:3306->3306/tcp, :::330
6->3306/tcp, 33060/tcp                                                                               53_mysql_1
2379b4802294   mher/flower:0.9.5                 "flower flower --bro…"   8 hours ago   Up 1 second   0.0.0.0:5555->5555/tcp, :::555
5->5555/tcp                                                                                          553_flower_1
27ef49d1f49e   rabbitmq:3.6-management-alpine    "docker-entrypoint.s…"   8 hours ago   Up 1 second   4369/tcp, 5671/tcp, 0.0.0.0:56
72->5672/tcp, :::5672->5672/tcp, 15671/tcp, 25672/tcp, 0.0.0.0:15672->15672/tcp, :::15672->15672/tcp  553_rabbitmq_1
```

接著安裝 API 所需套件

```
pipenv install fastapi==0.63.0 pandas==1.1.5 pymysql==1.0.2
sqlalchemy==1.4.20 uvicorn==0.14.0
```

以下是 API 範例，將資料從 MySQL 撈資料出來，藉由 API 回傳給使用者。

https://github.com/FinMind/FinMindBook/blob/master/DataEngineering/
Chapter6/6.3/6.3.2/main.py

main.py
```
import pandas as pd
from fastapi import FastAPI
from sqlalchemy import create_engine, engine
```

```python
def get_mysql_financialdata_conn() -> engine.base.Connection:
    address = "mysql+pymysql://root:test@127.0.0.1:3306/FinancialData"
    engine = create_engine(address)
    connect = engine.connect()
    return connect


app = FastAPI()


@app.get("/")
def read_root():
    return {"Hello": "World"}


@app.get("/taiwan_stock_price")
def taiwan_stock_price(
    stock_id: str = "",
    start_date: str = "",
    end_date: str = "",
):
    sql = f"""
    select * from taiwan_stock_price
    where StockID = '{stock_id}'
    and Date>= '{start_date}'
    and Date<= '{end_date}'
    """
    mysql_conn = get_mysql_financialdata_conn()
    data_df = pd.read_sql(sql, con=mysql_conn)
    data_dict = data_df.to_dict("records")
    return {"data": data_dict}
```

以上 API 的設計，讓使用者可以輸入股票代碼 stock_id、開始日期 start_date、結束日期 end_date，之後 API 接收到這些參數後，再去資料庫撈資料，回傳給使用者。

同樣，模擬使用者，對 API 發 request，結果如下，使用者成功拿到 data

啟動 API Server

```
pipenv run uvicorn main:app --reload --port 8888
```

模擬 user

```
sam@DESKTOP-IKT69L5:~/FinMindBook/DataEngineering/Chapter6/6.3/6.3.2$ python
Python 3.6.3 |Anaconda, Inc.| (default, Oct 13 2017, 12:02:49)
[GCC 7.2.0] on linux
Type "help", "copyright", "credits" or "license" for more information.
>>> import requests
>>> import pandas as pd
>>>
>>> payload = dict(
...     stock_id="2330",
...     start_date="2021-04-01",
...     end_date="2021-04-15",
... )
>>> res = requests.get("http://127.0.0.1:8888/taiwan_stock_price", params=payload)
>>> df = pd.DataFrame(res.json()["data"])
>>> df
  StockID  TradeVolume  Transaction    TradeValue   Open    Max    Min  Close  Change        Date
0    2330     45972766        48170   27520742963  598.0  602.0  594.0  602.0    15.0  2021-04-01
1    2330     37664216        42422   23045132094  615.0  616.0  608.0  610.0     8.0  2021-04-06
2    2330     28140964        28395   17175949080  614.0  614.0  608.0  610.0     0.0  2021-04-07
3    2330     26658283        32098   16217570357  606.0  613.0  603.0  613.0     3.0  2021-04-08
4    2330     31601619        28920   19344167081  618.0  618.0  609.0  610.0    -3.0  2021-04-09
>>>
```

到這步驟，API 基本上就完成第一階段了，是不是很簡單阿，接下來只要把 api 部屬在雲端機器上，使用者就能根據雲端 IP，對你的 API 發送 request，拿到資料。

那實際上 FastAPI 效能，到底有沒有比 Flask 好呢？以下章節將使用 ApacheBench 這個壓力測試的工具，對 API 做測試。

6.3.3 ApacheBench 壓力測試

根據官方文件介紹：

> *ab is a tool for benchmarking your Apache Hypertext Transfer Protocol (HTTP) server. It is designed to give you an impression of how your current Apache installation performs. This especially shows you how many requests per second your Apache installation is capable of serving.*

簡單來說呢，ApacheBench（簡稱 ab），是一個對伺服器進行基準測試的工具，可以模擬多個使用者，同時對伺服器發送多個 request，最後統計每次 request 需要多少時間，是否中間有 failed，以下將做個範例，分別對 Flask、FastAPI 做壓測，而 Flask、FastAPI 皆使用官方 GitHub 推薦的方法做啟動。

安裝

```
apt-get install apache2-utils -y
```

壓測指令如下：

```
ab -c 10 -n 1000  'http://127.0.0.1:8888/'
```

首先對 Flask 做壓測，結果如下，只看幾個重要指標就好。

```
Complete requests:      1000
Failed requests:        0
Total transferred:      164000 bytes
HTML transferred:       18000 bytes
Requests per second:    476.31 [#/sec] (mean)
Time per request:       20.995 [ms] (mean)
Time per request:       2.099 [ms] (mean, across all concurrent requests)
Transfer rate:          76.28 [Kbytes/sec] received
```

接著對 FastAPI 做壓測，結果如下：

```
Complete requests:      1000
Failed requests:        0
Total transferred:      142000 bytes
HTML transferred:       17000 bytes
Requests per second:    789.24 [#/sec] (mean)
Time per request:       12.670 [ms] (mean)
Time per request:       1.267 [ms] (mean, across all concurrent requests)
Transfer rate:          109.44 [Kbytes/sec] received
```

首先，介紹幾個統計指標：

Complete requests 代表完成幾個 request，這裡代表 1000 個都完成。

Failed requests 代表有幾個 request 失敗，如果你的伺服器不穩定，在高壓情況下，可能發生 1% 的失敗，這時 ApacheBench 就能很好的模擬出情境。

Total transferred 代表總共傳輸量，這裡 FastAPI 比 Flask 好約 20%。

Requests per second 代表每秒能承受多少 request。

Time per request 這裡有兩個，以 FastAPI 為例，分別是 12.670 ms、1.267 ms，差了 10 倍，剛好對應到，模擬 10 個使用者，第一個 12.670 ms，是這 10 個 user 的平均回應時間，第二個 1.267 ms，是每一個 user 的平均回應時間。FastAPI 跟 Flask 比較，快了 40%。

以上只是初步比較，當然 FastAPI 還有架構設計上的優點，這邊先不展示，且 FastAPI 有額外的功能，自動產生 Swagger 文件，這非常好用，下個章節將展示此功能。

6.3.4 API 介面 --- Swagger

Swagger 是一種自動生成的 OpenAPI 文件，可以直接在 Web 上對 API 發送 request，就不用寫 code 了，這對於開發上，有很大的幫助，先來看前面 FastAPI 自動生成的文件，到 http://127.0.0.1:8888/docs 以上網址，會顯示以下畫面：

這就是前面使用 FastAPI 架設所有的 API，因為前面章節，只設定兩種 API，因此只顯示兩種，接著來試試看用在網頁上，對 API 發送 request。

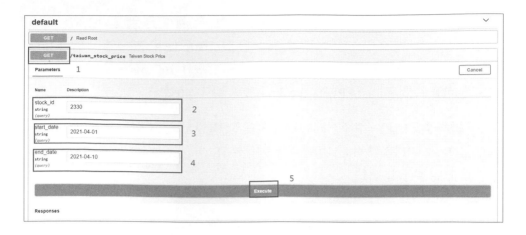

第一，先點 taiwan_stock_price 的 GET，之後會展開出輸入頁面，包含 stock_id、start_date、end_date 這三個參數，這也就是 API 程式中，定義需要輸入的參數，本書使用的框架，直接將程式碼轉換在 Web 介面。輸入完之後，再點 Execute 按鈕，就會發送 request 了，結果如下：

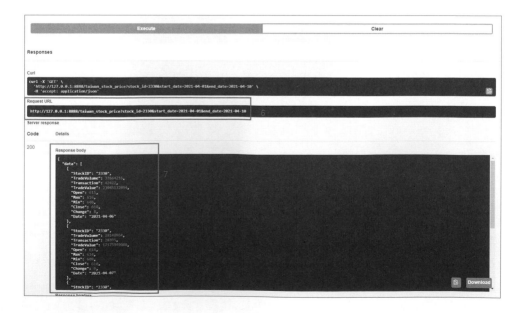

Request URL 的部分，就是網頁發送的 URL，包含帶的參數，而 Response body，就是 API 回應的結果，可以看到有股價相關的資料，這時相對於使用程式，有網站就能夠輕易的操作 API，非常方便，這也是 FastAPI 相較於 Flask 的優點之一。

本書到此，已經能架設 API 了，剩下的只需要將程式在雲端 Linode 上執行，修改一些設定，把 localhost、127.0.0.1 改成實體 IP，就能架設一個 API 讓別人來使用你的服務了，甚至還有 Swagger 介面，讓使用者清楚知道你的 API 有哪些功能。

目前爬蟲、API 已經完成第一步了，剩下只要持續新增爬蟲、API，就能讓資料越來越豐富，下個章節，將使用 Docker 包裝爬蟲、API。

Docker 主要用於後續持續開發的爬蟲、API，方便持續新版本更新迭代，試想一個情境，當未來 API 使用超過 2 台機器做分散，爬蟲也使用 2 台機器做分散，這時就有 4 台機器了，那更新一次爬蟲、API，需要連線到 4 台機器上，非常麻煩。而照著本書的方法，後續只需要按一個按鈕，就可以自動更新，當然自動化前期，需要花一點時間做開發，以下將一一介紹整個過程。

Chapter

07

容器管理工具 Docker

7.1 為什麼要用 Docker

前面章節的 MySQL、RabbitMQ 都用 Docker，那為什麼爬蟲跟 API 也需要用 Docker 呢？因為實務開發上，可能遇到以下幾個情境。

1. **開發環境與生產環境不完全相同**。在產品開發上，這是很常見的情境，團隊多個成員，可能同時用 Mac、Windows、Linux 做開發，雲端基本上都是 Linux 環境，因此開發的程式碼，在本地端可以順利運行，不保證在雲端上能正常，即使開發上也是 Linux，Linux 也有多種版本，而 Docker 是獨立環境，不論在 Windows、Linux、Mac，都不受影響，因此 Docker 可以確保開發與生產的穩定性。

2. **版本更動**。在開發過程中，會持續新增爬蟲、API，因此生產環境的版本，也會不斷更動，特別使用分散式架構，如果有 10 台機器，那換版過程將非常複雜，使用 Docker，再加上 Chapter 9 的 CI/CD，可以做到一鍵換版。

3. **拓展性**。本書使用分散式架構爬蟲，目前約 3-5 台機器，未來可能30-50 台，因此需考量拓展性，再加上 API、資料庫未來也會分散式，用 Loading Balance 做分流，因此需要可高拓展的架構，Docker就是高拓展性架構的基底。

4. **多服務、大型架構**。以數據分析來説，除了 API、爬蟲，將分析結果視覺化，是很常見的，因此需要 Web 才能展現給多數人，雖然Python 有 Web 的視覺化工具可以使用（如以 plotly 為基底的 dash），但高可用的 Web 基本上會走前後端分離，因此持續開發，除了爬蟲、API 外，還會需要許多服務。以 FinMind 為例，有爬蟲、API、分析、Backend、Web、DataBase、Airflow、Log 等多個服務，因此都用 Docker，可以很容易地跟其他服務做對接。

5. **K8s**。以現代架構來説，K8s 算是必備，那要如何學習 K8s？可以先從 Docker 開始，讓目前的服務容器化、微服務化，未來架構走到K8s，相對來説容易。

Docker 大致上有以上幾點好處，主要都是考量未來的架構設計，這也是筆者真實的經驗，真實體會過 Docker 的好處。下圖是 FinMind 目前的產品架構，一開始也是從爬蟲、API 開始，慢慢衍生出許多服務，而基本上整個產品都圍繞在 Docker 上，這也是為什麼本書一開始就先介紹Docker，Docker 幫筆者省去非常多時間。

以初學者來説，一開始開發產品時，可能不會接觸到這些，這需要花費不少時間踩雷，累積經驗，即使已經工作一段時間，由於公司分工細，你很難接觸到全貌，無法學習到某些技能，這是優點也是缺點，即使有接觸架構，可能也是接續前人的基礎，不過最難的還是一個真實場景，高流量會遇到完全不同的問題，也會考慮高效的設計架構，而本書直接提供以 FinMind 產品為例的架構，可以做初步參考，讓讀者在工作之餘，有額外管道學習這些技能。

FinMind 專注在多種金融數據上，同樣的方式，讀者可以嘗試在不同領域的數據上實現。以下章節將著重在架構層面上，關於爬蟲、API 的程式碼，在上面的章節已做過介紹，剩下只需花時間持續開發。

7.2 建立第一個 Docker Image--Dockerfile

7.2.1 爬蟲

架構如下：

```
├──── Dockerfile
├──── Makefile
├──── Pipfile
├──── Pipfile.lock
├──── README.md
├──── cache.Dockerfile
├──── crawler.yml
├──── crawler_multi_celery.yml
├──── create_partition_table.sql
├──── docker-compose.yml
├──── financialdata
│    ├──── __init__.py
│    ├──── backend
│    │    ├──── __init__.py
│    │    └──── db
│    │         ├──── __init__.py
│    │         ├──── clients.py
│    │         ├──── db.py
│    │         └──── router.py
│    ├──── config.py
│    ├──── crawler
│    │    ├──── __init__.py
│    │    ├──── taiwan_futures_daily.py
│    │    └──── taiwan_stock_price.py
│    ├──── producer.py
│    ├──── scheduler.py
│    ├──── schema
│    │    ├──── __init__.py
│    │    └──── dataset.py
│    └──── tasks
```

```
|        ├──── __init__.py
|        ├──── task.py
|        └──── worker.py
├──── genenv.py
├──── local.ini
├──── mysql.yml
├──── rabbitmq.yml
├──── scheduler.yml
└──── setup.py
```

要建立 Docker Image，基本上都是使用 Dockerfile 這個文件，首先，將爬蟲的程式碼，包成 Docker Image，使用以下 Dockerfile

https://github.com/FinMind/FinMindBook/blob/master/DataEngineering/Chapter7/7.2/7.2.1/Dockerfile

```Dockerfile
FROM continuumio/miniconda3:4.3.27

RUN apt-get update

RUN mkdir /FinMindProject
COPY . /FinMindProject/
WORKDIR /FinMindProject/

# install package
RUN pip install pipenv && pipenv sync

# genenv
RUN VERSION=RELEASE python genenv.py
```

先介紹幾個常用指令。

From。以 Image 為基底，在這上面去建立你的 Docker 環境。

Run。執行設定的指令，你可以使用此指令，去安裝所需的工具。

COPY。將本地端的檔案，複製進 Docker Image 裡面。

WORKDIR。設定你的工作目錄。

接著來一步步解析 Dockerfile 為什麼這樣寫：

1. 使用 miniconda 為基底，由於前面使用 miniconda，因此選用此 Image 作為基底。

```
FROM continuumio/miniconda3:4.3.27
```

2. 更新環境。

```
RUN apt-get update
```

3. 建立資料夾，待會存放爬蟲的程式碼。

```
RUN mkdir /FinMindProject
```

4. 將當前目錄下的檔案，複製進去，這裡使用點 . ，代表當前路徑所有檔案，複製進去 FinMindProject 這個資料夾。

```
COPY . /FinMindProject/
```

5. 設定工作目錄在 FinMindProject 底下，因為需要啟動 Celery 做爬蟲，因此設定在這目錄底下。

```
WORKDIR /FinMindProject/
```

6. 安裝 pipenv 且安裝所需 Package。

```
RUN pip install pipenv && pipenv sync
```

7. 建立環境變數。

```
RUN VERSION=RELEASE python genenv.py
```

以上指令就能能一步步安裝出所需的 Docker 環境,而初學遇到最大的問題是,不知如何寫起,這時可以先用以下指令 run docker,

```
docker run -it --rm continuumio/miniconda3:4.3.27 bash
```

成功後會進入到 Docker container 容器內

```
sam@DESKTOP-IKT69L5:~/FinMindBook$ docker run -it --rm continuumio/miniconda3:4.3.27 bash
root@b6db68387924:/#
```

再一步步嘗試指令,並記錄在 Dockerfile,最後找出 SOP,根據 SOP 流程,就能建立起所需環境。

而 Dockerfile 建立好了之後,使用以下指令,建立 Docker Image

```
docker build -f Dockerfile -t crawler:7.2.1 .
```

-f Dockerfile,是指定使用 Dockerfile 這個檔案,去建立 Image,如果你使用其他檔名,就在這更動檔名。

-t crawler:7.2.1,是指建立出的 Image 名稱。

.,就是當前目錄。

建立完 Docker Image 之後,使用 **Docker Images** 指令,就會多一個你剛剛建立的 Image:

```
sam@DESKTOP-IKT69L5:~/FinMindBook/DataEngineering/Chapter7/7.2/7.2.1$ docker images
REPOSITORY              TAG              IMAGE ID          CREATED           SIZE
crawler                 7.2.1            c5abc9f89a5b      16 seconds ago    866MB
```

這時有了爬蟲的 Docker Image 後,接著來用 Docker 啟動 Celery,而啟動方法與啟動 MySQL、RabbitMQ 的方法相同,Docker-Compose,以下是爬蟲的 Docker-Compose

https://github.com/FinMind/FinMindBook/blob/master/DataEngineering/
Chapter7/7.2/7.2.1/docker-compose.yml

```yml
docker-compose.yml

version: '3.0'
services:
  crawler_twse:
    image: crawler:7.2.1
    hostname: "twse"
    command: pipenv run celery -A financialdata.tasks.worker worker
--loglevel=info --concurrency=1  --hostname=%h -Q twse
    restart: always
    environment:
      - TZ=Asia/Taipei
    networks:
        - dev

networks:
  dev:
```

使用以下指令

```
docker-compose -f docker-compose.yml up
```

這時會出現以下錯誤

> *(2003, "Can't connect to MySQL server on '127.0.0.1' ([Errno 111] Connection refused)")*

你應該會很疑惑，你的電腦中，明明有啟動 MySQL，也可以透過 127.0.0.1 去做連線，為什麼會顯示無法連線？這是因為，Docker 內部是獨立環境，連 Network 網路也走獨立網路，那這時要如何讓 Docker 網路相通呢？這時需要先建立 Docker Network，指令如下

```
docker network create my_network
```

建立了 my_network Docker 網路，接著要額外設定，MySQL、RabbitMQ 都使用這個網路，需要修改 Docker-Compose 設定

https://github.com/FinMind/FinMindBook/blob/master/DataEngineering/
Chapter7/7.2/7.2.1/mysql.yml

```
mysql.yml
version: '3'
services:

  mysql:
    image: mysql:8.0
    command: mysqld --default-authentication-plugin=mysql_native_password
    ports:
      - 3306:3306
    environment:
      MYSQL_DATABASE: mydb
      MYSQL_USER: user
      MYSQL_PASSWORD: test
      MYSQL_ROOT_PASSWORD: test
    volumes:
      - mysql:/var/lib/mysql
    networks:
      - my_network
```

```
phpmyadmin:
    image: phpmyadmin/phpmyadmin:5.1.0
    links:
        - mysql:db
    ports:
        - 8000:80
    depends_on:
      - mysql
    networks:
        - my_network

networks:
  my_network:
    # 加入已經存在的網路
    external: true

volumes:
    mysql:
        external: true
```

這邊 Network 改為 my_network，並使用 external: true，讓 Docker-Compose 在啟動時，選擇加入現成的網路 my_network，RabbitMQ 也是同樣動作

https://github.com/FinMind/FinMindBook/blob/master/DataEngineering/Chapter7/7.2/7.2.1/rabbitmq.yml

rabbitmq.yml

```
version: '3'
services:

  rabbitmq:
    image: 'rabbitmq:3.6-management-alpine'
```

```
    ports:
      - '5672:5672'
      - '15672:15672'
    environment:
      RABBITMQ_DEFAULT_USER: "worker"
      RABBITMQ_DEFAULT_PASS: "worker"
      RABBITMQ_DEFAULT_VHOST: "/"
    networks:
      - my_network

  flower:
    image: mher/flower:0.9.5
    command: ["flower", "--broker=amqp://worker:worker@rabbitmq",
"--port=5555"]
    ports:
      - 5555:5555
    depends_on:
      - rabbitmq
    networks:
      - my_network

networks:
  my_network:
    # 加入已經存在的網路
    external: true
```

這時先關閉之前啟動的 MySQL、RabbitMQ。

先使用

```
docker ps
```

查看目前啟動的 Docker：

```
sam@DESKTOP-IKT69L5:~/FinMindBook/DataEnginner/Chapter5/5.5/5.5.5$ docker ps
CONTAINER ID   IMAGE                            COMMAND                CREATED          STATUS            PORTS
                                                                                        NAMES
c70c63dd2b06   phpmyadmin/phpmyadmin:5.1.0      "/docker-entrypoint..."   8 seconds ago   Up 6 seconds     0.0.0.0:8000->80/tcp, :::8000->80/tcp
                                                                                        555_phpmyadmin_1
06a555086c7b   mysql:8.0                        "docker-entrypoint.s…"    9 seconds ago   Up 7 seconds     0.0.0.0:3306->3306/tcp, :::3306->3306/tcp, 33060/
tcp                                                                                     555_mysql_1
0ad16ebb2ca9   mher/flower:0.9.5                "flower flower --bro…"   26 seconds ago   Up 23 seconds    0.0.0.0:5555->5555/tcp, :::5555->5555/tcp
                                                                                        555_flower_1
f5f422f299e8   rabbitmq:3.6-management-alpine   "docker-entrypoint.s…"   28 seconds ago   Up 26 seconds    4369/tcp, 5671/tcp, 0.0.0.0:5672->5672/tcp, :::56
72->5672/tcp, 15671/tcp, 25672/tcp, 0.0.0.0:15672->15672/tcp, :::15672->15672/tcp       555_rabbitmq_1
```

使用 docker stop 關閉 Container：

```
docker stop 555_flower_1 555_mysql_1 555_phpmyadmin_1 555_rabbitmq_1
```

使用 docker rm 刪除 Container：

```
docker rm 555_flower_1 555_mysql_1 555_phpmyadmin_1 555_rabbitmq_1
```

接著啟動新的 MySQL、RabbitMQ：

```
docker-compose -f rabbitmq.yml up -d
docker-compose -f mysql.yml up -d
```

這時爬蟲需要修改 local.init 設定，才能成功在 Docker Network 成功連上
MySQL、RabbitMQ，如下（單純擷取 RELEASE 部分）。

local.ini

```
[RELEASE]
MYSQL_DATA_HOST = mysql
MYSQL_DATA_USER = root
MYSQL_DATA_PASSWORD = test
MYSQL_DATA_PORT = 3306
MYSQL_DATA_DATABASE = FinancialData
WORKER_ACCOUNT = worker
WORKER_PASSWORD = worker
```

```
MESSAGE_QUEUE_HOST = rabbitmq
MESSAGE_QUEUE_PORT = 5672
```

特別注意 HOST 部分，在 Docker 內部，不是用 IP 做連線，而是使用 service name，以目前 mysql、rabbitmq 來說，service name 分別是 mysql、rabbitmq，如下圖：

用 service name 有幾點好處：

1. 你的 Docker 架構，可以很輕易的部屬在不同環境上，包含測試與產品環境上，不受 IP 限制。

2. 如果機器臨時需要維護，某些服務需要切換機器，如果用 IP，你所有服務都要改 IP，如果用 service name，那就不用改，方便很多。

3. 安全性，在多人開發中，如果 local.ini 包含實體 IP，你無法保證機器資訊會不會洩漏，因此使用 service name，即使對外洩漏，對方也無法知道，你的服務在哪台機器，只會知道，團隊使用的 MySQL host 是 mysql。

4. 拓展性，Loading Balance 分流，不過初期可能感受不到，也需要做額外設定，這裡先不提起，避免主題混淆。

修改完設定後，重新建立（build）Docker Image，並將 Docker-Compose 中的 networks 設定成 my_network：

https://github.com/FinMind/FinMindBook/blob/master/DataEngineering/ Chapter7/7.2/7.2.1/crawler.yml

```
crawler.yml
version: '3.0'
services:
  crawler_twse:
    image: crawler:7.2.1
    hostname: "twse"
    command: pipenv run celery -A financialdata.tasks.worker worker
--loglevel=info --concurrency=1  --hostname=%h -Q twse
    restart: always
    environment:
      - TZ=Asia/Taipei
    networks:
      - my_network

networks:
  my_network:
    # 加入已經存在的網路
    external: true
```

接著啟動 Celery

```
docker-compose -f crawler.yml up
```

如果結果如下圖，代表成功了：

```
sam@DESKTOP-IKT69L5:~/FinMindBook/DataEngineering/Chapter7/7.2/7.2.1$ docker-compose -f crawler.yml up
WARNING: Found orphan containers (721_mysql_1, 721_phpmyadmin_1, 721_rabbitmq_1, 721_flower_1) for this project. If you removed or
renamed this service in your compose file, you can run this command with the --remove-orphans flag to clean it up.
Starting 721_crawler_twse_1 ... done
Attaching to 721_crawler_twse_1
crawler_twse_1  | Loading .env environment variables...
crawler_twse_1  |
crawler_twse_1  |  -------------- celery@twse v5.1.0 (sun-harmonics)
crawler_twse_1  | --- ***** -----
crawler_twse_1  | -- ******* ---- Linux-5.4.72-microsoft-standard-WSL2-x86_64-with-debian-8.5 2021-10-31 00:44:31
crawler_twse_1  | - *** --- * ---
crawler_twse_1  | - ** ---------- [config]
crawler_twse_1  | - ** ---------- .> app:         task:0x7f3733aeef60
crawler_twse_1  | - ** ---------- .> transport:   amqp://worker:**@rabbitmq:5672//
crawler_twse_1  | - ** ---------- .> results:     disabled://
crawler_twse_1  | - *** --- * ---- .> concurrency: 1 (prefork)
crawler_twse_1  | -- ******* ---- .> task events: OFF (enable -E to monitor tasks in this worker)
crawler_twse_1  | --- ***** -----
crawler_twse_1  |  -------------- [queues]
crawler_twse_1  |                 .> twse          exchange=twse(direct) key=twse
crawler_twse_1  |
crawler_twse_1  | [tasks]
crawler_twse_1  |   . financialdata.tasks.task.crawler
```

可以看到，**Connected to amqp://worker:**@rabbitmq:5672//**，RabbitMQ 連線成功，並且連線的 host 是 RabbitMQ，也就是 service name，代表成功加入到 my_network，是在同一個 Docker 網路中。

這時跳轉到 http://localhost:5555/

可以看到 Celery 成功連線上，不過這只有 twse，接著再把其他 Celery 加上去：

https://github.com/FinMind/FinMindBook/blob/master/DataEngineering/Chapter7/7.2/7.2.1/crawler_multi_celery.yml

crawler_multi_celery.yml

```
version: '3.0'
services:
  crawler_twse:
    image: crawler:7.2.1
    hostname: "twse"
    command: pipenv run celery -A financialdata.tasks.worker worker
--loglevel=info --concurrency=1  --hostname=%h -Q twse
    restart: always
    environment:
      - TZ=Asia/Taipei
    networks:
        - my_network


  crawler_tpex:
    image: crawler:7.2.1
    hostname: "tpex"
    command: pipenv run celery -A financialdata.tasks.worker worker
--loglevel=info --concurrency=1  --hostname=%h -Q tpex
    restart: always
    environment:
      - TZ=Asia/Taipei
    networks:
        - my_network


  crawler_taifex:
    image: crawler:7.2.1
    hostname: "taifex"
    command: pipenv run celery -A financialdata.tasks.worker worker
--loglevel=info --concurrency=1  --hostname=%h -Q taifex
    restart: always
    environment:
      - TZ=Asia/Taipei
```

```
    networks:
        - my_network

networks:
  my_network:
    # 加入已經存在的網路
    external: true
```

接著啟動 Celery

```
docker-compose -f crawler_multi_celery.yml up
```

切換到 http://localhost:5555/，就可以看到，有三個 Celery

這時成功將爬蟲程式，包成 Docker Image，並成功連上 RabbitMQ、MySQL。

下一步，將 APScheduler 排程，也用 Docker 部屬，yml 檔如下：

https://github.com/FinMind/FinMindBook/blob/master/DataEngineering/Chapter7/7.2/7.2.1/scheduler.yml

scheduler.yml
```
version: '3.0'
services:
  scheduler:
    image: crawler:7.2.1
```

```
    hostname: "twse"
    command: pipenv run python financialdata/scheduler.py
    restart: always
    environment:
      - TZ=Asia/Taipei
    networks:
        - my_network


networks:
  my_network:
    # 加入已經存在的網路
    external: true
```

使用以下指令啟動 scheduler docker container

```
docker-compose -f scheduler.yml up
```

成功執行的畫面如下，這時時間到，就會自動發送 task 到 RabbitMQ 上，Worker 接收到任務，就會去執行爬蟲了。

```
sam@DESKTOP-IKT69L5:~/FinMindBook/DataEngineering/Chapter7/7.2/7.2.1$ docker-compose -f scheduler.yml up
WARNING: Found orphan containers (721_phpmyadmin_1, 721_crawler_twse_1, 721_rabbitmq_1, 721_mysql_1, 721_flower_1) for this project
. If you removed or renamed this service in your compose file, you can run this command with the --remove-orphans flag to clean it
up.
Creating 721_scheduler_1 ... done
Attaching to 721_scheduler_1
scheduler_1  | Loading .env environment variables...
scheduler_1  | 2021-10-31 00:45:46.504 | INFO     | __main__:main:27 - sent_crawler_task
```

當然如果你想讓他在背景執行，就執行以下指令

```
docker-compose -f scheduler.yml up -d
```

不用擔心壞掉怎麼辦，在 scheduler.yml 中有設定，

```
restart: always
```

壞了會自動重啟。

下一章節，將 API 也包成 Docker Image。

7.2.2 API

架構如下：

```
├──── api
│    ├──── config.py
│    ├──── __init__.py
│    └──── main.py
├──── api.yml
├──── Dockerfile
├──── genenv.py
├──── local.ini
├──── Makefile
├──── Pipfile
├──── Pipfile.lock
├──── README.md
│     setup.py
```

基本上與爬蟲是類似的，使用以下 Dockerfile 建立 Image：

https://github.com/FinMind/FinMindBook/blob/master/DataEngineering/
Chapter7/7.2/7.2.2/Dockerfile

```
Dockerfile
FROM continuumio/miniconda3:4.3.27

RUN apt-get update

RUN mkdir /FinMindProject
COPY . /FinMindProject/
WORKDIR /FinMindProject/
```

```
# install package
RUN pip install pipenv && pipenv sync

# genenv
RUN VERSION=RELEASE python genenv.py

# 預設執行的指令
CMD ["pipenv", "run", "uvicorn", "api.main:app", "--host", "0.0.0.0",
"--port", "8888"]
```

Dockerfile 與爬蟲的 SOP 是一樣的，先安裝好 Python 所需套件、環境變數後，比較特別的是，使用 CMD 設定預設執行的指令，這樣當 Run Docker Image 時，就會直接啟動 API，但為什麼在爬蟲時，不這樣寫呢？因為爬蟲使用 Celery，需要多種不同的 command，如 crawler_twse、crawler_tpex，因此 CMD 需要在 Docker-Compose 上設定，而 API 只有一種，因此寫在 Dockerfile 裡面，你也可以嘗試模仿爬蟲，將 CMD 寫在 Docker-Compose 中。

--host 0.0.0.0，是指定 API 不再用 local 當作 host，而是使用本機上的 IP 做為 host，原先不寫，預設是 localhost，但只限開發使用，如果要實際部屬，需額外設定是 0.0.0.0。

先使用以下指令，建立 Docker Image

```
docker build -f Dockerfile -t api:7.2.2 .
```

建立好 Image 之後，以下是 Docker-Compose：

https://github.com/FinMind/FinMindBook/blob/master/DataEngineering/Chapter7/7.2/7.2.2/api.yml

api.yml

```yaml
version: '3.0'
services:
  api:
    image: api:7.2.2
    ports:
        - 8888:8888
    hostname: "api"
    restart: always
    environment:
      - TZ=Asia/Taipei
    networks:
        - my_network

networks:
  my_network:
    # 加入已經存在的網路
    external: true
```

使用以下指令啟動 API

```
docker-compose -f api.yml up
```

接著如果看到以下畫面，就代表成功了

```
sam@DESKTOP-IKT69L5:~/FinMindBook/DataEngineering/Chapter7/7.2/7.2.2$ docker-compose -f api.yml up
Creating 722_api_1 ... done
Attaching to 722_api_1
api_1  | Loading .env environment variables...
api_1  | INFO:     Started server process [7]
api_1  | INFO:     Waiting for application startup.
api_1  | INFO:     Application startup complete.
api_1  | INFO:     Uvicorn running on http://0.0.0.0:8888 (Press CTRL+C to quit)
```

打開 http://127.0.0.1:8888/docs 此連結，會看到以下畫面

代表成功使用 Docker，架設好 API 了。

這時真實情境是，在開發機上做開發，並測試好用 Docker 架設 API，一切正常後，可以直接將 API Image，上傳到正式機上，並用 api.yml 啟動 Docker 部屬 API，並且不受開發機、正式機不同環境的限制了。

本書到此，成功使用 Docker 架設爬蟲、資料庫、RabbitMQ、API 以上服務了，同時也建立了專屬的爬蟲、API Docker Image，那要如何管理這些 Image 呢？以下將介紹 Docker Hub https://hub.docker.com/，Docker 官方網站，可以協助使用者，管理自己的 Image。

7.3 發布 Docker Image

7.3.1 Docker Hub 介紹

https://hub.docker.com/，Docker Hub，類似於 GitHub，GitHub 的 Repo 有 Public、Private 兩個方案，Public 是對外開源用途，Private 是私人開發用，Docker Hub 也一樣，有 Public、Private，例如想做一個開源服務，可以直接發布 Public 的 Docker Image，讓其他人直接下載，就像本書使用的 MySQL、RabbitMQ 這兩種服務，也是使用別人 Public 的 Image。

而 Private 可以存放不想公開的 Image，如爬蟲、API，在 local 開發好程式，並 Build 成 Image 後，上傳 Docker Hub，再到正式機，拉 Image 下來部屬。

7.3.2 發布 Docker Image

首先，請先到 Docker Hub 註冊，註冊後，先用 Docker 登入，（以下使用者名稱，用 linsamtw 為例，你需要更改為自己的）。

```
docker login --username linsamtw
```

接著再輸入密碼，登入成功後，會顯示以下畫面：

```
sam@DESKTOP-IKT69L5:~/FinMindBook/DataEnginner/Chapter6/6.3/6.3.2$ docker login --username linsamtw
Password:
WARNING! Your password will be stored unencrypted in /home/sam/.docker/config.json.
Configure a credential helper to remove this warning. See
https://docs.docker.com/engine/reference/commandline/login/#credentials-store

Login Succeeded
```

上傳 Image 前,需要對 Image 下 tag,如下:

```
docker tag crawler:7.2.1 linsamtw/crawler:7.2.1
```

上傳爬蟲的 Docker Image:

```
docker push linsamtw/crawler:7.2.1
```

同樣,也上傳 API 的 Docker Image:

```
docker tag api:7.2.2 linsamtw/api:7.2.2
docker push linsamtw/api:7.2.2
```

這時上去 Docker Hub,就可以看到剛剛發布的 Image,以下是本書上傳的 Image。

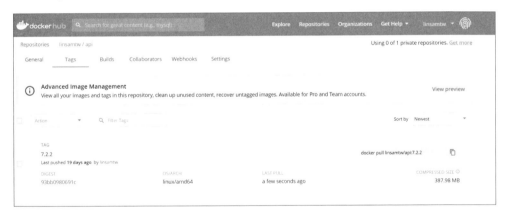

那在 Local Build 好 Image 後,要如何在雲端正式機部屬呢?

7.4 雲端部屬

先到雲端正式機，使用 docker login 指令登入後，使用 pull 指令，將發布的 Image 拉下來

```
docker pull linsamtw/api:7.2.2
```

```
sam@localhost:~/FinMindBook/DataEngineering/Chapter7/7.2/7.2.2$ docker pull linsamtw/api:7.2.2
7.2.2: Pulling from linsamtw/api
8ad8b3f87b37: Already exists
66b9b7852117: Already exists
4823bea101d0: Already exists
76d893c0ed7f: Already exists
188b30775d2f: Pull complete
95688c554cf2: Pull complete
5a85af52400a: Pull complete
4f4fb700ef54: Pull complete
9728154404c8: Pull complete
481fd4d0b191: Pull complete
Digest: sha256:99ed3d3bf8a12a45a84c5d4d8241dd60067f71912e83e418c04bffbfffb4beee
Status: Downloaded newer image for linsamtw/api:7.2.2
docker.io/linsamtw/api:7.2.2
sam@localhost:~/FinMindBook/DataEngineering/Chapter7/7.2/7.2.2$ []
```

接下來，可以直接 Run 這個 Image，去部屬你的 API，

```
docker run -it --rm -p 8888:8888 linsamtw/api:7.2.2
```

由於啟動 API 指令，寫在 Dockerfile 裡，因此直接 Run Image，就部屬好 API 了。

```
sam@localhost:~/FinMindBook/DataEngineering/Chapter7/7.2/7.2.2$ docker run -it --rm -p 8888:8888 linsamtw/api:7.2.2
Loading .env environment variables...
INFO:     Started server process [7]
INFO:     Waiting for application startup.
INFO:     Application startup complete.
INFO:     Uvicorn running on http://0.0.0.0:8888 (Press CTRL+C to quit)
```

那要如何在背景執行呢？總不可能一直開著這個 Terminal，只需要在啟動 Docker Image 時，多加一個指令 -d 即可，如下：

```
docker run -d -it --rm -p 8888:8888 linsamtw/api:7.2.2
```

這時就在正式機，部屬好 API 了。

同樣的，在 7.2，使用 Docker 在 local 建立起 RabbitMQ、MySQL、爬
蟲，這時模仿部屬 API，在雲端部屬以上服務，就可以在雲端，真正地部
屬整個產品環境了，因此整個開發流程是，

1. Local 開發完，測試沒問題。
2. 在 Local Build 成 Image，Push 至 Docker Hub。
3. 至雲端機器 Pull Image，部屬。

這樣能確保程式上，不受環境影響，且也使用 config.py、local.ini，去控
管開發與正式環境的設定，這能讓開發上，輕易切換不同環境。

到目前為止，部屬了爬蟲、API、RabbitMQ、MySQL，4 種不同的服
務，都用 Docker 啟動，那有沒有什麼好的管理工具，且能統一管理、更
新、查看所有 Log 並做監控呢？總不可能去一台台雲端上，查看 Docker
Log，要換版時，總不可能上去一台台的機器更換 Docker Image，如果有
10 台機器，那這樣也太笨了。

當然你可以自己寫程式做，但這問題我們不是第一個遇到，秉持著不要
自己造輪子的想法，將介紹 Docker Swarm 這套架構。

可能有些讀者會疑問，為什麼不選用 K8s。網路上蠻多 Swarm vs K8s 的
文章，Dwarm 主要是簡單、容易學習使用，K8s 多數用於大型架構，
且大多不會自架，而是直接使用 GCP 等雲服務，專案剛起步，先選用
Swarm，相對來説，開發比較快速，且 GCP 也不便宜，大多都只有公司
玩得起，還在學習階段，先用 Swarm 比較親民。

7.5 Docker Swarm

7.5.1 介紹

本書到現在，介紹了 Docker，用於啟動單一服務。Docker-Compose，
管理多個 Docker。如果遇到多個 Docker-Compose 時，要如何管理呢？
剛好在本書情境，就有 API、爬蟲、RabbitMQ、MySQL 這幾個 Docker-
Compose，這時就要使用 Docker Swarm 來進行管理，如下圖。

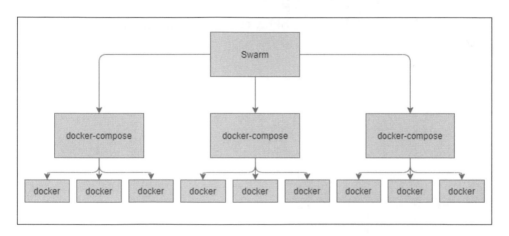

另外，Swarm 可以管理**多台機器**，這邊引入一個概念，**Manager** 和
Worker。在多台機器架構下，你會有一個負責作管控的 Manager 機器，
剩下其他的機器都是 Worker，Manager 有 Swarm 的管理權限，例如部屬
線上服務、更新線上服務、指定 Worker 執行特定任務等等，而 Worker
就只是單純做 Manager 分配好的工作，包含建立 MySQL、RabbitMQ、
API、爬蟲等，如以下情境。

我有 A、B 兩台機器，Manager 在 A 機器上 Worker 在 B 機器上，要部屬 MySQL，那要怎麼做呢？

首先，在 Manager 下達指令，要部屬 MySQL 這個服務在 Swarm 中，並指定在 Worker B 上跑 MySQL 這個 Docker Container，這時 B 接收到指令，並會自己去抓取相對應的 MySQL Image，並啟動這個 Image。

流程大致上就是這樣，管理權限全部掌握在 Manager 手上，所有的事，你直接在 Manager 下指令即可，他會傳達給 Worker，而 Worker 只是工人，負責跑 Manager 決定的工作，基本上不會直接操作到 Worker。

實際狀況如下，將 MySQL、RabbitMQ、API、Crawle 分別跑在不同機器上。

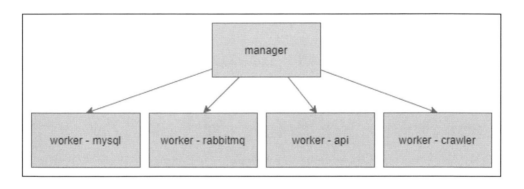

這樣的好處是，全部的 Services 都統一管理，相較於連去一台台的機器部屬 MySQL、RabbitMQ 等，甚至調整 MySQL、RabbitMQ 的 config 設定檔，我只要在 Manager 上，就可以管理所有的機器。

下個章節，將示範如何建立第一個 Swarm，並讓 Node 與 Manager 做連結。

7.5.2 建立第一個 swarm

首先，初始化 Docker Swarm

```
docker swarm init
```

畫面如下：

```
sam@localhost:~$ docker swarm init
Swarm initialized: current node (w5m87thetvi8e8kwifzkxscfm) is now a manager.

To add a worker to this swarm, run the following command:

    docker swarm join --token SWMTKN-1-1loirfpq5uyajclzor35t1ws4xt1z4hhwh141fn23vg5ntr58g-0p85mm8epx3ba5c3qqrtdlfr1 139.162.104.54:2377

To add a manager to this swarm, run 'docker swarm join-token manager' and follow the instructions.

sam@localhost:~$ 
```

這邊要特別注意，139.162.104.54 是 Linode 的實體 IP，如果你用個人電腦架設 Swarm，使用虛擬 IP，那其他電腦是找不到你的 IP 位址的，一般家庭跟電信商如中華電信，租用的光纖網路，大多都只有虛擬（浮動）IP，實體 IP 還需要額外設定，因此用 Linode 方便很多。

接著注意初始化後的一段訊息，上面寫

> *To add a worker to this swarm, run the following command:*
>
> *docker swarm join --token SWMTKN-1-1loirfpq5uyajclzor35t1 ws4xt1z4hhwh141fn23vg5ntr58g-0p85mm8epx3ba5c3qqrtdlfr1 139.162.104.54:2377*
>
> *To add a manager to this swarm, run 'docker swarm join-token manager' and follow the instructions.*

也就是說，如果你要加一個 Worker 到 Swarm，使用以下指令：

```
docker swarm join --token SWMTKN-1-1loirfpq5uyajclzor35t1ws4xt1z4hhwh141
fn23vg5ntr58g-0p85mm8epx3ba5c3qqrtdlfr1 139.162.104.54:2377
```

你可能會問，這段指令太長，需要額外記下來嗎？不用，在 Manager 的
機器上，使用以下指令，就會再次出現：

```
docker swarm join-token worker
```

```
sam@localhost:~$ docker swarm join-token worker
To add a worker to this swarm, run the following command:

    docker swarm join --token SWMTKN-1-1loirfpq5uyajclzor35t1ws4xt1z4hhwh141fn23vg5ntr58g-0p85mm8epx3ba5c3qqrtdlfr1 139.162.104.54:2377
```

這時就初始化成功了，只要在其他電腦上，執行以上指令，就會加入
Docker Wwarm 中，但沒有一個 UI 介面，怎麼知道目前 Swarm 下，有哪
些服務或是 Worker 呢？下一章節，將介紹 Portainer，Swarm 的 UI 介面
之一。

7.5.3 管理介面 --- Portainer

以下是 Portainer 的畫面，以 FinMind 為例，Swarm 內包含 57 個 Services、
221 個 Containers，管理多種服務，所以 Swarm 的上限是很高的，以下將
提供範例，一步步建立出你自己的 Portainer、Swarm 架構。

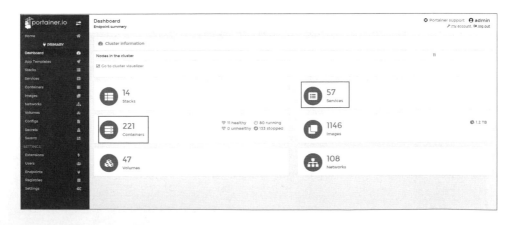

首先，一樣用 Docker 架設 Portainer，使用以下 Docker-Compose：

https://github.com/FinMind/FinMindBook/blob/master/DataEngineering/
Chapter7/7.5/7.5.3/portainer.yml

```yml
portainer.yml
version: '3.2'

services:
  agent:
    image: portainer/agent
    environment:
      AGENT_CLUSTER_ADDR: tasks.agent
    volumes:
      - /var/run/docker.sock:/var/run/docker.sock
      - /var/lib/docker/volumes:/var/lib/docker/volumes
    networks:
      - agent_network
    deploy:
      mode: global
      placement:
        constraints: [node.platform.os == linux]

  portainer:
    image: portainer/portainer:1.24.2
    command: -H tcp://tasks.agent:9001 --tlsskipverify
    ports:
      - "9000:9000"
      - "8000:8000"
    volumes:
      - portainer_data:/data
    networks:
      - agent_network
    deploy:
      mode: replicated
      replicas: 1
      placement:
```

```
        constraints: [node.role == manager]

networks:
  agent_network:
    driver: overlay

volumes:
  portainer_data:
```

在 Swarm 架構下，不再是使用 Docker-Compose 啟動，而是改用以下指令

```
docker stack deploy -c portainer.yml por
```

在解釋指令之前，先來看 Portainer 畫面，會比較好理解，進入以下連結，IP 需換成你自己 Linode 的 IP：

http://139.162.104.54:9000/

會看到以下畫面：

一開始，需要先設定帳號密碼，設定完後，按 Create user，之後會進入以下畫面，點選下面那隻鯨魚：

接著跳轉到以下畫面，再點選 Service：

跳轉到此畫面：

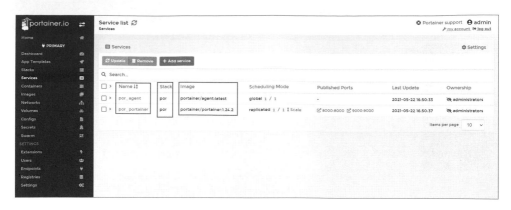

這時就出現剛剛用 portainer.yml 架設的 Container，先來介紹幾個名詞，
Name 底下的 por_agent、por_portainer，分別是上面 portainer.yml 中的兩個
services name，agent 和 portainer，如下，前面的 por，就是 Stack Name，

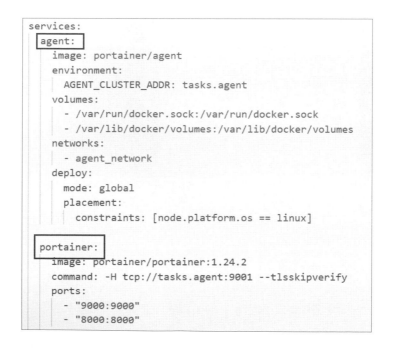

這時出現一個新名詞，Stack，英文翻譯是堆，如下圖，以目前範例來說，Stack 底下包含 agent、portainer，Stack 是一包、一堆的概念，例如 stack - por 這一堆，下面包含 agent、portainer 這兩個 Container，那為什麼要分 Stack 呢？因為後續包含更新服務、刪除服務等，都是以 Stack 為單位做操作。

接下來，將以上用 Docker-Compose 部屬的服務，全部改到 Swarm 上。初期會花許多時間在架構上，因為目前服務少，所以感受不到好處，可能還會想說，不如直接用 Docker-Compose，但是考量到做 Data 專案，不只是爬蟲、資料庫、API，還需要 Web 呈現視覺化、分析結果、紀錄 Log 等等，未來需要更多服務，因此用 Swarm 做管理，在未來拓展新服務時，可以直接在 Manager 操作，方便很多。

7.6 部屬服務

本章節，要部屬 MySQL、RabbitMQ、爬蟲、API 到 Swarm，架構如下：

```
├── Makefile
├── README.md
├── api
├── financialdata
├── mysql.yml
└── rabbitmq.yml
```

MySQL、RabbitMQ 分別使用 yml 就能部屬，api 跟 financialdata，需要建立 Docker Image，這兩個是獨立的 Project，建議都開新的 Project 做版控，FinMind 使用的是 GitLab，主要是因為要使用 GitLab-CI 進行後續的 CICD 自動化測試、部屬。

7.6.1 MySQL

將 MySQL 架設到 Swarm 架構中，跟 5.3.2 一樣，先在雲端機器上，建立 MySQL Volume

```
docker volume create mysql
```

建立 Network

```
docker network create --scope=swarm --driver=overlay my_network
```

如果發生 *network with name my_network already exists* 的問題，請先使用以下指令，刪除 my_network，如無法刪除，請先使用 docker stop container_name 指令，停止相關 Container，再刪除 my_network

```
docker network rm my_network
```

這裡跟之前比起來，比較特別，需要指定 **scope=swarm**，且 driver 指定 **overlay**，scope 指的就是 **network** 範圍，這裡用 **swarm**，而 driver 設為 **overlay**，是讓 swarm 內的服務可以互通，只要在 swarm 內部的機器，不論 5 台 10 台，都能互相通信。

接著使用一樣的 mysql.yml，並加上 Swarm 設定，如下：

https://github.com/FinMind/FinMindBook/blob/master/DataEngineering/Chapter7/7.6/mysql.yml

mysql.yml

```
version: '3'
services:

  mysql:
      image: mysql:8.0
      command: mysqld --default-authentication-plugin=mysql_native_password
      ports:
          - 3306:3306
      environment:
          MYSQL_DATABASE: mydb
          MYSQL_USER: user
          MYSQL_PASSWORD: test
          MYSQL_ROOT_PASSWORD: test
      volumes:
          - mysql:/var/lib/mysql
      # swarm 設定
      deploy:
        mode: replicated
        replicas: 1
        placement:
          constraints: [node.labels.mysql == true]
      networks:
          - my_network

  phpmyadmin:
      image: phpmyadmin/phpmyadmin:5.1.0
      links:
          - mysql:db
      ports:
          - 8080:80
      environment:
          MYSQL_USER: user
          MYSQL_PASSWORD: test
```

```
            MYSQL_ROOT_PASSWORD: test
            PMA_HOST: mysql
        depends_on:
          - mysql
        # swarm 設定
        deploy:
          mode: replicated
          replicas: 1
          placement:
            constraints: [node.labels.mysql == true]
        networks:
            - my_network

networks:
  my_network:
    # 加入已經存在的網路
    external: true


volumes:
  mysql:
    external: true
```

這裡有兩點需要特別注意，

第一，**PMA_HOST**

```
    environment:
        MYSQL_USER: user
        MYSQL_PASSWORD: test
        MYSQL_ROOT_PASSWORD: test
        PMA_HOST: mysql
```

在 phpmyadmin 要額外設定 environment，特別是 PMA_HOST，在 Swarm
架構下，需要指定 MySQL 的 host 位址，這也是跟 Docker-Compose 不太

一樣的地方，一開始如果直接轉到 Swarm 下，會遇到許多坑，本書已經
踩過，並提供解法。

第二，**labels**

```
constraints: [node.labels.mysql == true]
```

這是指定，要在特定 node 上跑 MySQL Services，Node 就等同於 Worker，
例如我有 3 台機器，要怎麼指定特定服務如 MySQL，在不同機器上執
行，就是靠這裡的設定。

在部屬 MySQL 之前，需要先到 Portainer 上做設定 Node 的 Labels，步驟
如下：

1. 到 Portainer 頁面，點選 **Swarm**。

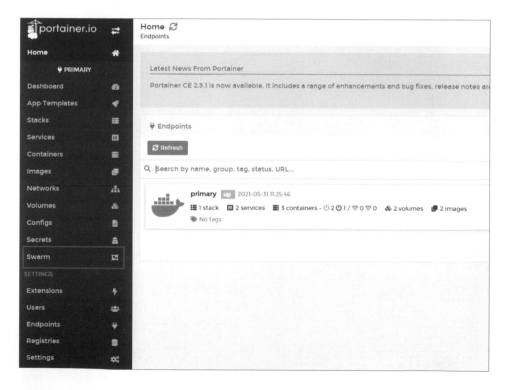

2. 點選 Nodes 底下的 **test**，這裡我是指定 test 這個 Worker 來架設 MySQL，如果你有多台機器，可以選其他機器。

3. 接著點選 **Label**，會新增一個 Label，在 **Name** 的地方，輸入 **MySQL**，在 **Value** 的地方，輸入 **true**，最後點選 **Apply changes**，就成功加上 Label 了。

Host overview 🔁
Docker

</> Host Details

Hostname	test
OS Information	linux x86_64
Total CPU	1
Total memory	1 GB

</> Engine Details

Version	18.09.7
Volume Plugins	local
Network Plugins	bridge, host, macvlan, null, overlay

</> Node Details

Role	manager (172.105.219.195:2377)
Availability	Active ▾
Status	ready
Node Labels	⊕ label **1**

Label		Value	
Name mysql **2**		Value true **3**	

Apply changes ▾ **4**

4. 有 Label 之後，使用以下指令，這是專門用來部屬 Swarm 的指令。

```
docker stack deploy --with-registry-auth -c mysql.yml mysql
```

5. 這時候在 Swarm 上，就可以看到 MySQL 的 Services

6. **preparing** 是這個 Services 正在準備中的狀態，等 Docker 成功 Run 起來後，會改成以下狀態，**running**。

這時成功部屬好 MySQL，到 http://139.162.104.54:8080/（139.162.104.54 改成你機器的 IP），就可以看到 MySQL 了。

使用 Portainer 介面，就可以很清楚的知道，MySQL 的狀態，接著部屬 RabbitMQ。

7.6.2 RabbitMQ

使用的 Docker-Compose 如下，同樣需要做 Swarm 設定：

https://github.com/FinMind/FinMindBook/blob/master/DataEngineering/ Chapter7/7.6/rabbitmq.yml rabbitmq.yml

```yaml
version: '3'
services:

  rabbitmq:
    image: 'rabbitmq:3.6-management-alpine'
    ports:
      - '5672:5672'
      - '15672:15672'
    environment:
      RABBITMQ_DEFAULT_USER: "worker"
      RABBITMQ_DEFAULT_PASS: "worker"
      RABBITMQ_DEFAULT_VHOST: "/"
    # swarm 設定
    deploy:
      mode: replicated
      replicas: 1
      placement:
        constraints: [node.labels.rabbitmq == true]
    networks:
      - my_network

  flower:
    image: mher/flower:0.9.5
    command: ["flower", "--broker=amqp://worker:worker@rabbitmq",
"--port=5555"]
    ports:
      - 5555:5555
    depends_on:
      - rabbitmq
    # swarm 設定
```

```
  deploy:
    mode: replicated
    replicas: 1
    placement:
      constraints: [node.labels.flower == true]
  networks:
    - my_network


networks:
  my_network:
    # 加入已經存在的網路
    external: true
```

同樣，需要先到 Portainer 上，將 Labels 加上去，如下：

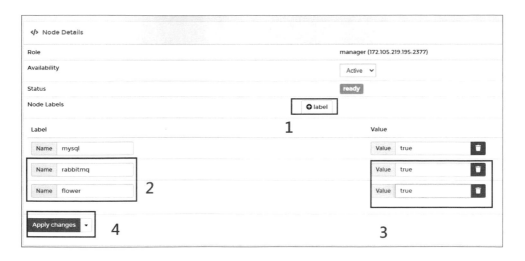

接著執行以下指令：

```
docker stack deploy --with-registry-auth -c rabbitmq.yml rabbitmq
```

這時在 Portainer 上，就能看到 RabbitMQ、Flower 這兩個 Services，如
下：

這時到 http://139.162.104.54:15672/（139.162.104.54 改成你機器的 IP），
就會看到 RabbitMQ 的畫面：

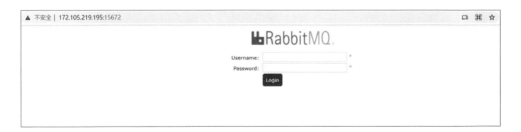

到 http://139.162.104.54:5555/（139.162.104.54 改成你機器的 IP），就會
看到 Flower 的畫面。

如果畫面都跟上面一樣，那就代表部屬成功了，本書到此，已經使用 Docker Swarm，同時管理 MySQL、Flower、RabbitMQ，下一步，把爬蟲也部屬上去。

7.6.3　爬蟲

首先，到雲端機器上，將章節 7.3.2 上傳 Docker Hub 的 Docker Image，Pull 下來：

```
docker pull linsamtw/crawler:7.2.1
```

當然你也可以自己重新 Build，指令就參考前面章節，接著使用以下 Docker-Compose 部屬在 Swarm 上：

https://github.com/FinMind/FinMindBook/blob/master/DataEngineering/ Chapter7/7.6/financialdata/crawler.yml

crawler.yml

```
version: '3.0'
services:
  crawler_twse:
    image: linsamtw/crawler:7.2.1
```

```
    hostname: "twse"
    command: pipenv run celery -A financialdata.tasks.worker worker
--loglevel=info --concurrency=1  --hostname=%h -Q twse
    restart: always
    # swarm 設定
    deploy:
      mode: replicated
      replicas: 1
      placement:
        constraints: [node.labels.crawler_twse == true]
    environment:
      - TZ=Asia/Taipei
    networks:
      - my_network

networks:
  my_network:
    # 加入已經存在的網路
    external: true
```

https://github.com/FinMind/FinMindBook/blob/master/DataEngineering/
Chapter7/7.6/financialdata/scheduler.yml

scheduler.yml

```
version: '3.0'
services:
  scheduler:
    image: linsamtw/crawler:7.2.1
    hostname: "twse"
    command: pipenv run python financialdata/scheduler.py
    restart: always
    # swarm 設定
```

```
    deploy:
      mode: replicated
      replicas: 1
      placement:
        constraints: [node.labels.crawler_scheduler == true]
      environment:
        - TZ=Asia/Taipei
      networks:
        - my_network


networks:
  my_network:
    # 加入已經存在的網路
    external: true
```

這裡同時部屬爬蟲跟 scheduler，跟上面相同，需要再次設定 Labels：

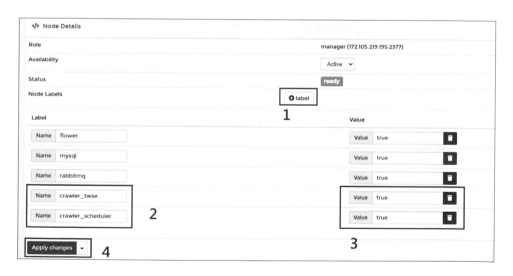

先部屬 crawler_twse 就好,之後很容易可以再拓展其他 Celery。因為 MySQL 是全新的,所以在部屬之前,需要先建立 DataBase,不然待會 Celery 的 Client,會連接不到 DataBase 報 Error。

將以下 SQL 複製到 MySQL 上並執行（此 SQL 包含建立 **FinancialData** 資料庫,因此不需再額外手動建立 **FinancialData** DataBase）。

https://github.com/FinMind/FinMindBook/blob/master/DataEngineering/ Chapter7/7.6/financialdata/create_partition_table.sql

```
create_partition_table.sql
CREATE DATABASE `FinancialData`;
CREATE TABLE `FinancialData`.`taiwan_stock_price`(
    `StockID` VARCHAR(10) NOT NULL,
    `TradeVolume` BIGINT NOT NULL,
    `Transaction` INT NOT NULL,
    `TradeValue` BIGINT NOT NULL,
    `Open` FLOAT NOT NULL,
    `Max` FLOAT NOT NULL,
    `Min` FLOAT NOT NULL,
    `Close` FLOAT NOT NULL,
    `Change` FLOAT NOT NULL,
    `Date` DATE NOT NULL,
    PRIMARY KEY(`StockID`, `Date`)
)
PARTITION BY RANGE(YEAR(Date)) (
    PARTITION p2005 VALUES LESS THAN (2006),
    PARTITION p2006 VALUES LESS THAN (2007),
    PARTITION p2007 VALUES LESS THAN (2008),
    PARTITION p2008 VALUES LESS THAN (2009),
    PARTITION p2009 VALUES LESS THAN (2010),
```

```
        PARTITION p2010 VALUES LESS THAN (2011),
        PARTITION p2011 VALUES LESS THAN (2012),
        PARTITION p2012 VALUES LESS THAN (2013),
        PARTITION p2013 VALUES LESS THAN (2014),
        PARTITION p2014 VALUES LESS THAN (2015),
        PARTITION p2015 VALUES LESS THAN (2016),
        PARTITION p2016 VALUES LESS THAN (2017),
        PARTITION p2017 VALUES LESS THAN (2018),
        PARTITION p2018 VALUES LESS THAN (2019),
        PARTITION p2019 VALUES LESS THAN (2020),
        PARTITION p2020 VALUES LESS THAN (2021),
        PARTITION p2021 VALUES LESS THAN (2022),
        PARTITION p2022 VALUES LESS THAN (2023),
        PARTITION p2023 VALUES LESS THAN (2024),
        PARTITION p2024 VALUES LESS THAN (2025)

);

CREATE TABLE `FinancialData`.`taiwan_futures_daily`(
    `Date` DATE NOT NULL,
    `FuturesID` VARCHAR(10) NOT NULL,
    `ContractDate` VARCHAR(30) NOT NULL,
    `Open` FLOAT NOT NULL,
    `Max` FLOAT NOT NULL,
    `Min` FLOAT NOT NULL,
    `Close` FLOAT NOT NULL,
    `Change` FLOAT NOT NULL,
    `ChangePer` FLOAT NOT NULL,
    `Volume` FLOAT NOT NULL,
    `SettlementPrice` FLOAT NOT NULL,
    `OpenInterest` INT NOT NULL,
```

```
    `TradingSession` VARCHAR(11) NOT NULL,
    PRIMARY KEY(`FuturesID`, `Date`)
)
PARTITION BY RANGE(YEAR(Date)) (
    PARTITION p2005 VALUES LESS THAN (2006),
    PARTITION p2006 VALUES LESS THAN (2007),
    PARTITION p2007 VALUES LESS THAN (2008),
    PARTITION p2008 VALUES LESS THAN (2009),
    PARTITION p2009 VALUES LESS THAN (2010),
    PARTITION p2010 VALUES LESS THAN (2011),
    PARTITION p2011 VALUES LESS THAN (2012),
    PARTITION p2012 VALUES LESS THAN (2013),
    PARTITION p2013 VALUES LESS THAN (2014),
    PARTITION p2014 VALUES LESS THAN (2015),
    PARTITION p2015 VALUES LESS THAN (2016),
    PARTITION p2016 VALUES LESS THAN (2017),
    PARTITION p2017 VALUES LESS THAN (2018),
    PARTITION p2018 VALUES LESS THAN (2019),
    PARTITION p2019 VALUES LESS THAN (2020),
    PARTITION p2020 VALUES LESS THAN (2021),
    PARTITION p2021 VALUES LESS THAN (2022),
    PARTITION p2022 VALUES LESS THAN (2023),
    PARTITION p2023 VALUES LESS THAN (2024),
    PARTITION p2024 VALUES LESS THAN (2025)
);
```

複製到以下地方：

接著，使用以下指令做部屬爬蟲：

```
docker stack deploy --with-registry-auth -c financialdata/crawler.yml
financialdata
```

這時到 Portainer，就會出現 financialdata_crawler_twse，代表成功部屬，如果想看 Celery log，可以直接點下面 Actions 第一個圖示：

跳轉到 Log 頁面後，Log 就顯示在以下畫面：

```
Container logs
Containers > financialdata_crawler_twse.1.ut6jlua3mwo296ljkzpvfg7c0 > Logs

📄 Log viewer settings

Auto-refresh logs ?  ⚫
Wrap lines  ⚫
Display timestamps  ⚪
Fetch        All logs                    ▼
Search       Filter...
Lines        100
Actions      📋 Copy   📋 Copy selected lines   ✕ Unselect

-------------- celery@twse v5.0.5 (singularity)
--- ***** -----
-- ******* ---- Linux-4.4.0-210-generic-x86_64-with-debian-8.5 2021-06-01 01:54:53
- *** --- * ---
- ** ---------- [config]
- ** ---------- .> app:         task:0x7f3b36819f60
- ** ---------- .> transport:   amqp://worker:**@rabbitmq:5672//
- ** ---------- .> results:     disabled://
- *** --- * --- .> concurrency: 1 (prefork)
-- ******* ---- .> task events: OFF (enable -E to monitor tasks in this worker)
--- ***** -----
-------------- [queues]
               .> twse          exchange=twse(direct) key=twse

[tasks]
  . financialdata.tasks.task.crawler
/root/.local/share/virtualenvs/FinMindProject-5Pr1mRrU/lib/python3.6/site-packages/celery/platforms.py:797: RuntimeWarning: You're running the worker with superuser privileges: this is
absolutely not recommended!
Please specify a different user using the --uid option.
User information: uid=0 euid=0 gid=0 egid=0
  uid=uid, euid=euid, gid=gid, egid=egid,
[2021-06-01 01:54:53,861: INFO/MainProcess] Connected to amqp://worker:**@rabbitmq:5672//
[2021-06-01 01:54:53,884: INFO/MainProcess] mingle: searching for neighbors
[2021-06-01 01:54:54,940: INFO/MainProcess] mingle: all alone
[2021-06-01 01:54:54,986: INFO/MainProcess] celery@twse ready.
[2021-06-01 01:54:58,240: INFO/MainProcess] Events of group {task} enabled by remote.
```

這是另一個好處，不用再去每台機器上收集、觀看 Log，Portainer 上面就有 Log 可以直接看，甚至有其他服務如 ELK，專門收集分散式 Log，這裡就不多作介紹，主要是想表達，Swarm 架構上，大部分的資訊都統一處理。機器少可能感覺不出來，等超過 10 台機器，這架構好處就很明顯了。

最後再部屬 scheduler：

```
docker stack deploy --with-registry-auth -c financialdata/scheduler.yml financialdata
```

執行以上指令，成功後，Portainer 上就會出現 financialdata_scheduler 的 Services 了。

到這步驟後，完成爬蟲部屬，並且有排程會每天抓新資料了，那歷史資料怎麼辦呢？

點選下圖的按鈕：

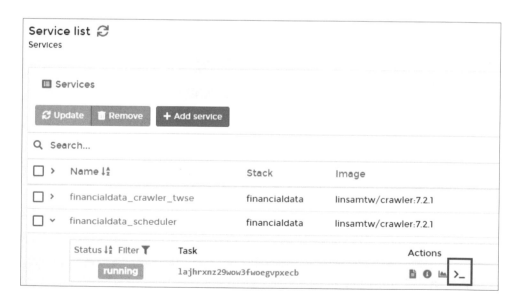

再點選 Connect：

Container console
Containers > financialdata_scheduler.1.lajhrxnz29wow3fwoegvpxecb > Console

>_ Execute

Command	⬡	/bin/bash

Use custom command ⬭

User ❓   root

[Connect]

就會進入到 Container 內部：

Container console
Containers > financialdata_scheduler.1.lajhrxnz29wow3fwoegvpxecb > Console

>_ Execute

Exec into container as default user **using command** bash [Disconnect]

```
root@twse:/FinMindProject# pipenv run python financialdata/producer.py taiwan_stock_price 2021-04-01 2021-04-12
```

這時就可以發送任務，抓歷史資料（日期可自行調整，這裡用 2021-04-01 到 2021-04-12 作為例子），指令如下：

```
pipenv run python financialdata/producer.py taiwan_stock_price 2021-04-01 2021-04-12
```

再去 RabbitMQ http://139.162.104.54:15672/#/queues（139.162.104.54 改成你機器的 IP）上查看 Queues 的狀態：

任務已經成功發送給 RabbitMQ，並且 twse 的 queue 已經被 Worker 執行了，但是 tpex 沒有，因為上面 Celery 只有啟動 twse 而已，這部分讀者可再自行調整。

那 Worker 執行任務,有沒有成功呢?到 http://139.162.104.54:5555/
(139.162.104.54 改成你機器的 IP)查看 Flower:

任務都成功了,再去 MySQL http://139.162.104.54:8080/(139.162.104.54
改成你機器的 IP)上查看資料是否正確地上傳:

資料也確實上傳了,下一步再部屬 API,整個產品就算完成第一步了。

7.6.4 API

部屬 API Services,同樣,以下是 yml 檔:

https://github.com/FinMind/FinMindBook/blob/master/DataEngineering/
Chapter7/7.6/api/api.yml

```
api.yml
version: '3.0'
services:
  api:
    image: linsamtw/api:7.2.2
    ports:
        - 8888:8888
    hostname: "api"
    restart: always
    # swarm 設定
    deploy:
      mode: replicated
      replicas: 1
      placement:
        constraints: [node.labels.api == true]
    environment:
      - TZ=Asia/Taipei
    networks:
        - my_network


networks:
  my_network:
    # 加入已經存在的網路
    external: true
```

設定 deploy，labels 設定 api=true，接著去 Portainer 上面，設定要執行 API 機器的 Label，這裡依然使用同一台機器。

Docker 非常省資源，一台 5 美金的 Linode，就能同時部屬目前用到的所有服務：

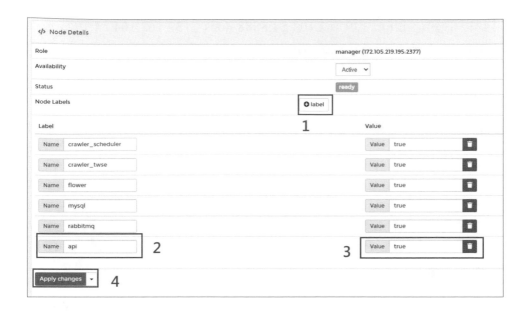

使用以下指令部屬 API

```
docker stack deploy --with-registry-auth -c api/api.yml api
```

接著再到 Portainer 上看，就會出現 API 了。

再到以下網址 http://139.162.104.54:8888/docs（139.162.104.54 改成你機器的 IP），就會看到成功部屬 API 了。

在 7.6.3，有成功爬蟲抓台股股價到資料庫，那試著對 API 發送 request，觀察是否拿的到資料，步驟如下：

發送 API，前面爬取到 2021-04-01 的資料，因此這裡 start_date、end_date 的區間選 2021-04-01、2021-04-10：

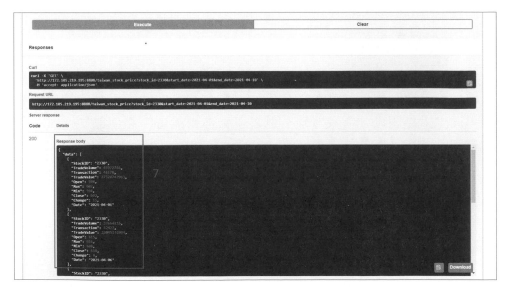

成功發送 API，且回傳 2330 股價資料，代表 API 成功。

本書到這，產品已經完成初版了，雖然目前只有台股股價，但整個流程已經串通，接下來只要照著架構持續開發新爬蟲，新 API，基本上不會有太大問題，且本書使用 Celery、RabbitMQ 分散式架構，因此效能問題不大，也不會被 ban IP，且雲端不怕機器壞掉，重點是，到現在，只使用 1~2 台機器而已，一個月只要 5~10 美金。

未來持續開發，架構慢慢變大，就會遇到幾個問題，**架構零散，導致重構困難，手動換版、手動測試，沒有固定流程，難以維護**，這也是團隊初期，最容易遇到的問題，急著開發新功能，導致欠缺測試、沒有 CICD 流程，造成換版常常掛掉，想重構卻容易把程式搞壞，難以維護等等的技術債，甚至最後砍掉重練（當然這是需要快速開發產品，做的取捨，本書單純探討工程方面的問題）。

為了解決以上問題，在 **Chapter 8、Chapter 9**，將介紹 **Unit Test、CICD**，讓產品在開發一個階段，確定方向後，開始建立良好的開發流程，雖然會花許多時間，但可以保證後續產品的穩定性，不用擔心程式有 bug、新版本掛掉等問題。

某些讀者可能會認為，程式有 bug，這是開發者的問題，但筆者認為，這一部分也是管理者的責任，如果先把自動化測試做上去，就能夠自動把關，減少人為因素，當只管理一個專案，你可能有時間手動做測試，但如果同時管理 3 個以上的專案呢？自動化可以省去很多時間，減少錯誤發生，何樂而不為呢？

第 2 篇
產品迭代 -- 測試運維

自動化測試

8.1　單元測試 Unit Test

8.1.1　什麼是單元測試 Unit Test？

在程式開發中，會將許多程式寫成 function 函數，讓程式模組化，但在長期開發中，一定會經歷過**重構**階段，重新設計架構，那這時會遇到一個問題，要如何確保，重構完後，程式結果不會出錯呢？通常對程式做大改，很容易出現錯誤，這時就要引入測試，確保程式結果不變。

程式架構中，函數是最小單元，而單元測試，就是最小化的測試。將所有的函數，都做一個對應的測試函數，確保每一個函數不會因為某次改動而壞掉。除此之外，測試在多人開發上，有一個很大的幫助，其他人在看你寫的程式時，不需要直接執行程式，單純從測試上，就能清楚知道每個函數的 intput、output。

口頭上說可能無法理解，以下舉幾個例子作為範例。

8.1.2　第一個測試

首先，介紹第一個測試，讓讀者做個最基本的了解。關於 Unit Test 測試架構，有多種框架，如 Python 原生支援的 **unittest**、第三方框架 **pytest**，本書選用 pytest，主因是，pytest 支持 unittest、且在 GitHub 得到 7.4k stars，算是主流架構。

測試範例架構如下。

```
├── Pipfile
├── Pipfile.lock
├── README.md
├── __init__.py
├── sample1.py
└── test_sample1.py
```

安裝 pytest

```
pipenv install pytest==6.2.4
```

sample1.py 就是主程式，或是可以看成你的函數，**test_sample1.py** 就是對應的測試，基本上測試的程式規範是，在要測試的 .py 前面，加上 **test**，

例如 **sample1.py**，那對應的測試，就命名為 **test_sample1.py**，而程式碼如下。

https://github.com/FinMind/FinMindBook/blob/master/DataEngineering/
Chapter8/8.1.2/sample1.py

```
sample1.py
def add(x, y):
    return x + y
```

https://github.com/FinMind/FinMindBook/blob/master/DataEngineering/
Chapter8/8.1.2/test_sample1.py

```
test_sample1.py
from sample1 import add


def test_add():
    result = add(1, 2)
    expected = 3
    assert result == expected
```

舉個最簡單的例子，在 **sample1.py** 寫一個 **add** 函數，做加總，而測試就是，執行 **add** 函數。命名規則是，原函數前面加上 **test**，如 **test_add**，然後再寫上 **expected**，預期執行 **add** 函數得到的結果，最後設定 **assert**，檢查 **add** 函數結果，是否如預期。

接著要如何執行測試呢？

有兩種方法，使用 VS Code 插鍵，或是用 Command Line 指令執行，以下將一一介紹這兩種方法。

VS Code 插鍵

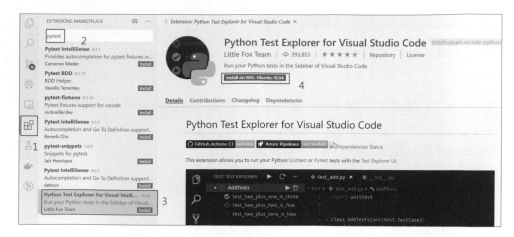

在 VS Code，根據以上步驟，安裝好 pytest 插鍵後，重啟 VS Code，之後會看到以下畫面：

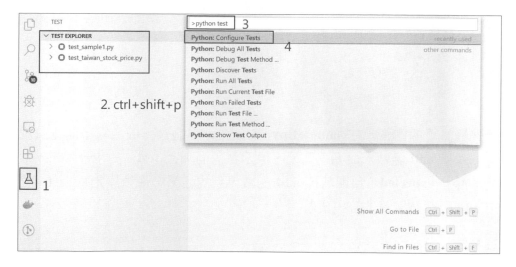

你的 VS Code IDE，左邊的欄位，會多一個藥水，這是 VS Code 提供的測試工具，但還沒有任何的測試，需要設定 VS Code，選用哪個 Python 的測試框架。

鍵盤輸入 **ctrl+shift+p**，搜尋 **python test**，選取 **Python: Configure**：

VS Code 會出現三個測試框架，選擇 pytest，接著 VS Code 需要選擇，要測試哪個目錄路徑，選取第一個選項 **.**，當前路徑：

設定完成，左邊畫面出現了測試：

接著本書將介紹，如何使用 GUI 工具，執行測試

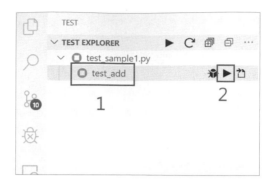

在測試的地方，可以看到，剛剛寫的 test_sample1.py，而底下的 test_add，就是測試函數，接著點擊箭頭（2），就會開始執行 test_add 的測試

藍色的 …，代表正在執行中測試，成功就會顯示綠色 ✓，

VS Code 插件，主要是在開發上，方便測試單一函數，在專案開發中，你可能有上百個測試，跑一次通常都需要花費不少時間，因此只測特定函數，是非常有幫助的。接下來介紹，使用 Command line 指令做測試。

command line

使用以下指令做測試

```
pipenv run pytest test_sample1.py
```

```
sam@DESKTOP-IKT69L5:~/FinMindBook/DataEngineering/Chapter7/7.2/7.2.1$ docker-compose -f scheduler.yml up
WARNING: Found orphan containers (721_phpmyadmin_1, 721_crawler_twse_1, 721_rabbitmq_1, 721_mysql_1, 721_flower_1) for this project
. If you removed or renamed this service in your compose file, you can run this command with the --remove-orphans flag to clean it
up.
Creating 721_scheduler_1 ... done
Attaching to 721_scheduler_1
scheduler_1  | Loading .env environment variables...
scheduler_1  | 2021-10-31 00:45:46.504 | INFO     | __main__:main:27 - sent_crawler_task
```

成功後，就會顯示上圖，這就代表測試成功了，那指令有什麼好處呢？在 Chapter 9，將介紹 CI 自動化測試，寫流程自動跑測試，一般在 Git 開發上，每次 Merge 前，都必須跑過測試，過了才能安心 Merge，因此需要指令做測試，而在 VS Code 上手動跑測試，不符合自動化的概念，VS Code 測試主要是開發上使用。

介紹完簡單的測試之後，將使用前面開發的爬蟲、API 專案做範例，介紹實務上的測試怎麼執行。

8.1.3 測試覆蓋率

在介紹爬蟲、API 測試範例之前，先介紹一個測試的進階概念，**測試覆蓋率（Test Coverage）**。顧名思義，這是判斷整個專案，被測試覆蓋的比例，舉例來說，你的專案有 10 個 function 函數，而你只寫了 3 個對應的測試，那你的覆蓋率就是，3/10，30%，代表你的測試還有很大的努力空間。

測試覆蓋率，當然是越高越好，但這邊要先提醒一件事，即使到了 100%
覆蓋率，依然不代表程式不會出錯，很多情況，是商業邏輯上的判斷問
題，而非程式本身的問題。

最後，測試不是萬能，但是沒有測試，是萬萬不能，覆蓋率越高，越能
保證你的程式穩定度。

8.1.4 爬蟲單元測試範例

爬蟲也可以寫測試嗎？筆者過去聽過一種說法，爬蟲因為是去抓取對方
網站，而對方網站會不穩定，所以寫爬蟲測試沒用。

這其實是一個不好的觀念，因為測試無法 100% 保證程式穩定性，所以就
不做嗎？不做就是 0 分，做了至少有 60 分，以下介紹爬蟲測試範例。

架構如下，tests 資料夾與 financialdata 在同一層級，tests/crawler/test_
taiwan_stock_price.py，這隻程式就是台股爬蟲的測試。

https://github.com/FinMind/FinMindBook/tree/master/DataEngineering/
Chapter8/8.1.4

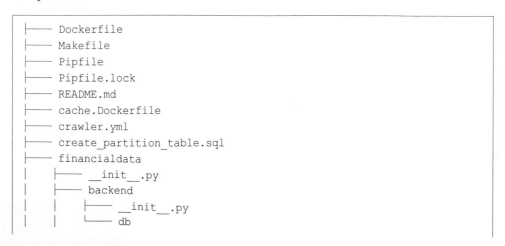

```
├── Dockerfile
├── Makefile
├── Pipfile
├── Pipfile.lock
├── README.md
├── cache.Dockerfile
├── crawler.yml
├── create_partition_table.sql
├── financialdata
│   ├── __init__.py
│   ├── backend
│   │   ├── __init__.py
│   │   └── db
```

```
│     │         ├───  __init__.py
│     │         ├───  clients.py
│     │         ├───  db.py
│     │         └───  router.py
│     ├───  config.py
│     ├───  crawler
│     │         ├───  __init__.py
│     │         ├───  taiwan_futures_daily.py
│     │         └───  taiwan_stock_price.py
│     ├───  producer.py
│     ├───  scheduler.py
│     ├───  schema
│     │         ├───  __init__.py
│     │         └───  dataset.py
│     └───  tasks
│               ├───  __init__.py
│               ├───  task.py
│               └───  worker.py
├───  genenv.py
├───  local.ini
├───  pytest.ini
├───  .coveragerc
├───  scheduler.yml
├───  setup.py
└───  tests
          ├───  __init__.py
          └───  crawler
                    ├───  __init__.py
                    └───  test_taiwan_stock_price.py
```

這裡新增了 pytest.ini、.coveragerc 這兩個關於測試的設定檔，主要是測試設定檔、測試覆蓋率設定檔，另外新增了兩個 Package，用於進階的測試與測試覆蓋率計算，安裝方法如下：

```
pipenv install pytest-cov==2.11.1 pytest-mock==3.5.1
```

安裝完之後，來看看測試範例，test_taiwan_stock_price.py 的覆蓋率，指令如下：

```
pipenv run pytest --cov-report term-missing --cov-config=.coveragerc
--cov=./financialdata/ tests/
```

測試結果如下圖：

框框 1，Cover，是每個主程式的測試覆蓋率，這個指令會自動計算覆蓋率，這裡先以 taiwan_stock_price.py 的測試作範例，因此其他的 Cover 都是 0。

框框 2，這裡 test_taiwan_stock_price.py 測試覆蓋率達到 100%，代表爬蟲程式中的各種情境，都被測試覆蓋到。

以上測試過了，測試率也到達 100%，接下來，將一步步介紹測試程式碼。

這裡需特別注意一點，主程式 taiwan_stock_price.py 只有 316 列，但 test_taiwan_stock_price.py 卻有 581 列，這也是為什麼大部分的人，不喜歡也缺少寫測試的習慣，因為在 coding 上，會花掉至少一半的開發時間。

由於專案初期開發上，會要求產品快速產出，因此缺少測試，是很合理的。但產品到一定階段後，必須將測試補上，不然整份專案將變成一塊磚，沒有人敢去動，因為非常容易把程式搞壞，後續的人更難維護。

https://github.com/FinMind/FinMindBook/blob/master/DataEngineering/Chapter8/8.1.4/tests/crawler/test_taiwan_stock_price.py

test_taiwan_stock_price.py，關於每個測試的情境、細節，本書都寫在註解上：

```python
import pandas as pd

from financialdata.crawler.taiwan_stock_price import (
    clear_data,
    colname_zh2en,
    convert_change,
    convert_date,
    crawler,
    gen_task_paramter_list,
    is_weekend,
    set_column,
    twse_header,
    tpex_header,
    crawler_twse,
    crawler_tpex,
)
from financialdata.schema.dataset import (
    check_schema,
)


def test_is_weekend_false():
```

```python
    """
    測試, 非周末, 輸入周一 1, 回傳 False
    """
    result = is_weekend(day=1)   # 執行結果
    expected = False
    # 先寫好預期結果, 這樣即使不執行程式,
    # 單純看測試, 也能了解這個程式的執行結果
    assert (
        result == expected
    )  # 檢查, 執行結果 == 預期結果


def test_is_weekend_true():
    """
    測試, 是周末, 輸入週日 0, 回傳 False
    """
    result = is_weekend(day=0)   # 執行結果
    expected = True
    # 先寫好預期結果, 這樣即使不執行程式,
    # 單純看測試, 也能了解這個程式的執行結果
    assert (
        result == expected
    )  # 檢查, 執行結果 == 預期結果


def test_gen_task_paramter_list():
    """
    測試建立 task 參數列表, 2021-01-01 ~ 2021-01-05
    """
    result = gen_task_paramter_list(
        start_date="2021-01-01",
        end_date="2021-01-05",
```

```
)  # 執行結果
expected = [
    {
        "date": "2021-01-01",
        "data_source": "twse",
    },
    {
        "date": "2021-01-01",
        "data_source": "tpex",
    },
    {
        "date": "2021-01-02",
        "data_source": "twse",
    },
    {
        "date": "2021-01-02",
        "data_source": "tpex",
    },
    {
        "date": "2021-01-05",
        "data_source": "twse",
    },
    {
        "date": "2021-01-05",
        "data_source": "tpex",
    },
]
# 預期得到 2021-01-01 ~ 2021-01-05 的任務參數列表
# 再發送這些參數到 rabbitmq, 給每個 worker 單獨執行爬蟲
assert (
    result == expected
)  # 檢查, 執行結果 == 預期結果
```

```python
def test_clear_data():
    # 準備好 input 的假資料
    df = pd.DataFrame(
        [
            {
                "StockID": "0050",
                "TradeVolume": "4,962,514",
                "Transaction": "6,179",
                "TradeValue": "616,480,760",
                "Open": "124.20",
                "Max": "124.65",
                "Min": "123.75",
                "Close": "124.60",
                "Change": 0.25,
                "Date": "2021-01-05",
            },
            {
                "StockID": "0051",
                "TradeVolume": "175,269",
                "Transaction": "44",
                "TradeValue": "7,827,387",
                "Open": "44.60",
                "Max": "44.74",
                "Min": "44.39",
                "Close": "44.64",
                "Change": 0.04,
                "Date": "2021-01-05",
            },
            {
                "StockID": "0052",
```

```
            "TradeVolume": "1,536,598",
            "Transaction": "673",
            "TradeValue": "172,232,526",
            "Open": "112.10",
            "Max": "112.90",
            "Min": "111.15",
            "Close": "112.90",
            "Change": 0.8,
            "Date": "2021-01-05",
        },
    ]
)
result_df = clear_data(
    df.copy()
)  # 輸入函數，得到結果
expected_df = pd.DataFrame(
    [
        {
            "StockID": "0050",
            "TradeVolume": "4962514",
            "Transaction": "6179",
            "TradeValue": "616480760",
            "Open": "124.20",
            "Max": "124.65",
            "Min": "123.75",
            "Close": "124.60",
            "Change": "0.25",
            "Date": "2021-01-05",
        },
        {
            "StockID": "0051",
            "TradeVolume": "175269",
```

```
                    "Transaction": "44",
                    "TradeValue": "7827387",
                    "Open": "44.60",
                    "Max": "44.74",
                    "Min": "44.39",
                    "Close": "44.64",
                    "Change": "0.04",
                    "Date": "2021-01-05",
            },
            {
                    "StockID": "0052",
                    "TradeVolume": "1536598",
                    "Transaction": "673",
                    "TradeValue": "172232526",
                    "Open": "112.10",
                    "Max": "112.90",
                    "Min": "111.15",
                    "Close": "112.90",
                    "Change": "0.8",
                    "Date": "2021-01-05",
            },
        ]
)
# 預期結果，做完資料清理
# 將原先的會計數字，如 1,536,598
# 轉換為一般數字 1536598
assert (
    pd.testing.assert_frame_equal(
        result_df, expected_df
    )
    is None
)  # 檢查，執行結果 == 預期結果
```

```python
def test_colname_zh2en():
    # 準備好 input 的假資料
    result_df = pd.DataFrame(
        [
            {
                0: "0050",
                1: "元大台灣50",
                2: "4,962,514",
                3: "6,179",
                4: "616,480,760",
                5: "124.20",
                6: "124.65",
                7: "123.75",
                8: "124.60",
                9: "<p style= color:red>+</p>",
                10: "0.25",
                11: "124.55",
                12: "123",
                13: "124.60",
                14: "29",
                15: "0.00",
            },
            {
                0: "0051",
                1: "元大中型100",
                2: "175,269",
                3: "44",
                4: "7,827,387",
                5: "44.60",
                6: "44.74",
```

```
                7: "44.39",
                8: "44.64",
                9: "<p style= color:red>+</p>",
                10: "0.04",
                11: "44.64",
                12: "20",
                13: "44.74",
                14: "2",
                15: "0.00",
            },
        ]
    )
colname = [
        "證券代號",
        "證券名稱",
        "成交股數",
        "成交筆數",
        "成交金額",
        "開盤價",
        "最高價",
        "最低價",
        "收盤價",
        "漲跌(+/-)",
        "漲跌價差",
        "最後揭示買價",
        "最後揭示買量",
        "最後揭示賣價",
        "最後揭示賣量",
        "本益比",
    ]
result_df = colname_zh2en(
        result_df.copy(), colname
```

```
)  # 輸入函數，得到結果
expected_df = pd.DataFrame(
    [
        {
            "StockID": "0050",
            "TradeVolume": "4,962,514",
            "Transaction": "6,179",
            "TradeValue": "616,480,760",
            "Open": "124.20",
            "Max": "124.65",
            "Min": "123.75",
            "Close": "124.60",
            "Dir": "<p style= color:red>+</p>",
            "Change": "0.25",
        },
        {
            "StockID": "0051",
            "TradeVolume": "175,269",
            "Transaction": "44",
            "TradeValue": "7,827,387",
            "Open": "44.60",
            "Max": "44.74",
            "Min": "44.39",
            "Close": "44.64",
            "Dir": "<p style= color:red>+</p>",
            "Change": "0.04",
        },
    ]
)
# 預期結果，將 raw data ，包含中文欄位，
# 轉換成英文欄位，以便存進資料庫
assert (
```

```python
        pd.testing.assert_frame_equal(
            result_df, expected_df
        )
        is None
    )  # 檢查, 執行結果 == 預期結果


def test_twse_header():
    result = twse_header()
    expected = {
        "Accept": "application/json, text/javascript, */*; q=0.01",
        "Accept-Encoding": "gzip, deflate",
        "Accept-Language": "zh-TW,zh;q=0.9,en-US;q=0.8,en;q=0.7",
        "Connection": "keep-alive",
        "Host": "www.twse.com.tw",
        "Referer": "https://www.twse.com.tw/zh/page/trading/exchange/
MI_INDEX.html",
        "User-Agent": "Mozilla/5.0 (Windows NT 10.0; Win64;
x64) AppleWebKit/537.36 (KHTML, like Gecko) Chrome/71.0.3578.98
Safari/537.36",
        "X-Requested-With": "XMLHttpRequest",
    }
    assert result == expected


def test_tpex_header():
    result = tpex_header()
    expected = {
        "Accept": "application/json, text/javascript, */*; q=0.01",
        "Accept-Encoding": "gzip, deflate",
        "Accept-Language": "zh-TW,zh;q=0.9,en-US;q=0.8,en;q=0.7",
        "Connection": "keep-alive",
```

```
        "Host": "www.tpex.org.tw",
        "Referer": "https://www.tpex.org.tw/web/stock/aftertrading/
otc_quotes_no1430/stk_wn1430.php?l=zh-tw",
        "User-Agent": "Mozilla/5.0 (Windows NT 10.0; Win64; x64)
AppleWebKit/537.36 (KHTML, like Gecko) Chrome/73.0.3683.103
Safari/537.36",
        "X-Requested-With": "XMLHttpRequest",
    }
    assert result == expected


def test_set_column():
    # 準備好 input 的假資料
    df = pd.DataFrame(
        [
            {
                0: "00679B",
                2: "44.91",
                3: "-0.08",
                4: "45.00",
                5: "45.00",
                6: "44.85",
                7: "270,000",
                8: "12,127,770",
                9: "147",
            },
            {
                0: "00687B",
                2: "47.03",
                3: "-0.09",
                4: "47.13",
                5: "47.13",
```

```
                6: "47.00",
                7: "429,000",
                8: "20,181,570",
                9: "39",
        },
        {
                0: "00694B",
                2: "37.77",
                3: "-0.07",
                4: "37.84",
                5: "37.84",
                6: "37.72",
                7: "343,000",
                8: "12,943,630",
                9: "35",
        },
    ]
)
result_df = set_column(
    df
)  # 輸入函數，得到結果
expected_df = pd.DataFrame(
    [
        {
                "StockID": "00679B",
                "Close": "44.91",
                "Change": "-0.08",
                "Open": "45.00",
                "Max": "45.00",
                "Min": "44.85",
                "TradeVolume": "270,000",
                "TradeValue": "12,127,770",
```

```
                    "Transaction": "147",
                },
                {
                    "StockID": "00687B",
                    "Close": "47.03",
                    "Change": "-0.09",
                    "Open": "47.13",
                    "Max": "47.13",
                    "Min": "47.00",
                    "TradeVolume": "429,000",
                    "TradeValue": "20,181,570",
                    "Transaction": "39",
                },
                {
                    "StockID": "00694B",
                    "Close": "37.77",
                    "Change": "-0.07",
                    "Open": "37.84",
                    "Max": "37.84",
                    "Min": "37.72",
                    "TradeVolume": "343,000",
                    "TradeValue": "12,943,630",
                    "Transaction": "35",
                },
            ]
        )
        # 預期結果，根據資料的位置，設置對應的欄位名稱
        assert (
            pd.testing.assert_frame_equal(
                result_df, expected_df
            )
            is None
```

```python
    )  # 檢查, 執行結果 == 預期結果


def test_crawler_twse_data9():
    """
    測試在證交所, 2021 正常爬到資料的情境,
    data 在 response 底下的 key, data9
    一般政府網站, 長時間的資料, 格式常常不一致
    """
    result_df = crawler_twse(
        date="2021-01-05"
    )  # 執行結果
    assert (
        len(result_df) == 20596
    )  # 檢查, 資料量是否正確
    assert list(result_df.columns) == [
        "StockID",
        "TradeVolume",
        "Transaction",
        "TradeValue",
        "Open",
        "Max",
        "Min",
        "Close",
        "Change",
        "Date",
    ]  # 檢查, 資料欄位是否正確


def test_crawler_twse_data8():
    """
    測試在證交所, 2008 正常爬到資料的情境, 時間不同, 資料格式不同
```

```
    data 在 response 底下的 key, data8
    一般政府網站, 長時間的資料, 格式常常不一致
    """
    result_df = crawler_twse(
        date="2008-01-04"
    )
    assert (
        len(result_df) == 2760
    )  # 檢查, 資料量是否正確
    assert list(result_df.columns) == [
        "StockID",
        "TradeVolume",
        "Transaction",
        "TradeValue",
        "Open",
        "Max",
        "Min",
        "Close",
        "Change",
        "Date",
    ]  # 檢查, 資料欄位是否正確


def test_crawler_twse_no_data():
    """
    測試沒 data 的時間點, 爬蟲是否正常
    """
    result_df = crawler_twse(
        date="2000-01-04"
    )
    assert (
        len(result_df) == 0
```

```
)  # 沒 data, 回傳 0
# 沒 data, 一樣要回傳 pd.DataFrame 型態
assert isinstance(
    result_df, pd.DataFrame
)


def test_crawler_twse_error(mocker):
    """
    測試, 情境為, 對方網站回傳例外狀況, 或是被 ban IP 時, 爬蟲是否會失敗

    這邊使用特別的技巧, mocker,
    因為在測試階段, 無法保證對方 server 一定會給錯誤的結果
    因此使用 mocker, 對 requests 做"替換", 換成設定的結果
    如下
    """
    # 將特定路徑下的 requests 替換掉
    mock_requests = mocker.patch(
        "financialdata.crawler.taiwan_stock_price.requests"
    )
    # 將 requests.get 的回傳值 response, 替換掉成 ""
    # 如此一來, 當在測試爬蟲時,
    # 發送 requests 得到的 response, 就會是 ""
    mock_requests.get.return_value = ""
    result_df = crawler_twse(
        date="2000-01-04"
    )
    assert (
        len(result_df) == 0
    )  # 沒 data, 回傳 0
    # 沒 data, 一樣要回傳 pd.DataFrame 型態
    assert isinstance(
```

```
        result_df, pd.DataFrame
    )


def test_crawler_tpex_success():
    """
    測試櫃買中心，爬蟲成功時的狀況
    """
    result_df = crawler_tpex(
        date="2021-01-05"
    )  # 執行結果
    assert (
        len(result_df) == 6609
    )  # 檢查，資料量是否正確
    assert list(result_df.columns) == [
        "StockID",
        "Close",
        "Change",
        "Open",
        "Max",
        "Min",
        "TradeVolume",
        "TradeValue",
        "Transaction",
        "Date",
    ]


def test_crawler_tpex_no_data():
    """
    測試沒 data 的時間點，爬蟲是否正常
    """
```

```python
    result_df = crawler_tpex(
        date="2021-01-01"
    )
    assert (
        len(result_df) == 0
    )  # 沒 data, 回傳 0
    # 沒 data, 一樣要回傳 pd.DataFrame 型態
    assert isinstance(
        result_df, pd.DataFrame
    )


def test_convert_change():
    # 準備好 input 的假資料
    df = pd.DataFrame(
        [
            {
                "StockID": "0050",
                "TradeVolume": "4,680,733",
                "Transaction": "5,327",
                "TradeValue": "649,025,587",
                "Open": "139.00",
                "Max": "139.20",
                "Min": "138.05",
                "Close": "138.30",
                "Dir": "<p style= color:green>-</p>",
                "Change": "0.65",
                "Date": "2021-07-01",
            },
            {
                "StockID": "0051",
                "TradeVolume": "175,374",
```

```
            "Transaction": "120",
            "TradeValue": "10,152,802",
            "Open": "58.20",
            "Max": "59.10",
            "Min": "57.40",
            "Close": "57.90",
            "Dir": "<p style= color:green>-</p>",
            "Change": "0.30",
            "Date": "2021-07-01",
        },
        {
            "StockID": "0052",
            "TradeVolume": "514,042",
            "Transaction": "270",
            "TradeValue": "64,127,738",
            "Open": "125.00",
            "Max": "125.20",
            "Min": "124.35",
            "Close": "124.35",
            "Dir": "<p style= color:green>-</p>",
            "Change": "0.65",
            "Date": "2021-07-01",
        },
    ]
)
result_df = convert_change(
    df
)  # 執行結果
expected_df = pd.DataFrame(
    [
        {
            "StockID": "0050",
```

```
        "TradeVolume": "4,680,733",
        "Transaction": "5,327",
        "TradeValue": "649,025,587",
        "Open": "139.00",
        "Max": "139.20",
        "Min": "138.05",
        "Close": "138.30",
        "Change": -0.65,
        "Date": "2021-07-01",
    },
    {
        "StockID": "0051",
        "TradeVolume": "175,374",
        "Transaction": "120",
        "TradeValue": "10,152,802",
        "Open": "58.20",
        "Max": "59.10",
        "Min": "57.40",
        "Close": "57.90",
        "Change": -0.3,
        "Date": "2021-07-01",
    },
    {
        "StockID": "0052",
        "TradeVolume": "514,042",
        "Transaction": "270",
        "TradeValue": "64,127,738",
        "Open": "125.00",
        "Max": "125.20",
        "Min": "124.35",
        "Close": "124.35",
        "Change": -0.65,
```

```
                "Date": "2021-07-01",
            },
        ]
    )
    # 預期結果,
    # 將 Dir (正負號) 與 Change (漲跌幅) 結合
    assert (
        pd.testing.assert_frame_equal(
            result_df, expected_df
        )
        is None
    )  # 檢查, 執行結果 == 預期結果


def test_convert_date():
    date = (
        "2021-07-01"  # 準備好 input 的假資料
    )
    result = convert_date(date)  # 執行結果
    expected = "110/07/01"  # 預期結果
    assert (
        result == expected
    )  # 檢查, 執行結果 == 預期結果


def test_crawler_twse():
    # 測試證交所爬蟲, end to end test
    result_df = crawler(
        parameter={
            "date": "2021-01-05",
            "data_source": "twse",
        }
```

```
    )
    result_df = check_schema(
        result_df, "TaiwanStockPrice"
    )
    assert len(result_df) > 0


def test_crawler_tpex():
    # 測試櫃買中心爬蟲, end to end test
    result_df = crawler(
        parameter={
            "date": "2021-01-05",
            "data_source": "tpex",
        }
    )
    result_df = check_schema(
        result_df, "TaiwanStockPrice"
    )
    assert len(result_df) > 0
```

這裡用到進階技巧，**mocker**，可參考上面 test_crawler_twse_error 函數，
該情境主要是因為在測試階段，需要發送 Requests，但無法保證對方回傳
的結果。

舉例來說，爬蟲的網站在半夜 12 點，會固定維護，當下發送 Requests
時，得到的結果與其他時間點不同，因此平常的時段，測不到這種情
境，那怎麼辦呢？

因應上面情境，使用 mocker 替換掉 requests，在測試階段，將 requests
回傳的結果寫定，固定成半夜 12 點回傳的結果，這樣可以保證在測試當
下，是模擬半夜 12 點的情境，並針對這個 Case，做例外處理。

當寫好測試後，可以觀察 8.1.2 中 VS Code 的插件功能，會顯示所有的測試，並可針對特定測試去執行

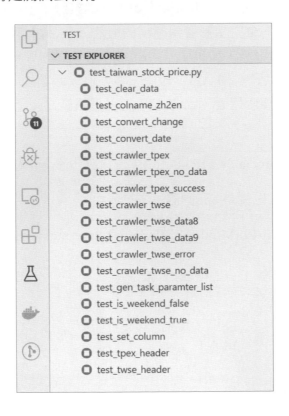

該爬蟲專案，只將 taiwan_stock_price.py 的測試覆蓋率做到 100%，其他部分，就留給讀者做後續的練習，只要增加測試後，持續執行以下指令

```
pipenv run pytest --cov-report term-missing --cov-config=.coveragerc
--cov=./financialdata/ tests/
```

就能看著測試覆蓋率慢慢提高，盡可能將覆蓋率到 80%、90% 以上，會是比較好的習慣。下一章節，將介紹 API 的測試。

8.1.5 API 單元測試範例

相對爬蟲來說，API 的測試更重要，如果爬蟲新版本上線後壞了，頂多當下做修正即可，不會直接影響到產品的服務。但 API 不同，因為 API 是直接面對 User 的服務，如果壞了，User 會直接受影響，不論是 Backend 的 API，或是提供資料的 API，即使沒壞，但改版前後，結構 Schema 不同，也會大大影響使用者，因此 API 必須更嚴謹，以下拿 7.6 的 API 來做測試範例。

架構如下，跟 financialdata 一樣，tests 資料夾也跟 api 在同一層。

tests/test_main.py 也就是 api 的測試。

https://github.com/FinMind/FinMindBook/tree/master/DataEngineering/Chapter8/8.1.5

```
├──── Dockerfile
├──── Makefile
├──── Pipfile
├──── Pipfile.lock
├──── README.md
├──── api
│    ├──── __init__.py
│    ├──── config.py
│    └──── main.py
├──── api.yml
├──── genenv.py
├──── local.ini
├──── pytest.ini
├──── .coveragerc
├──── setup.py
└──── tests
     ├──── __init__.py
     └──── test_main.py
```

由於 API 比較單純，因此測試也不會太複雜，是個方便作為入門的範例。程式如下：

同樣，先安裝測試 Package：

```
pipenv install pytest-cov==2.11.1 pytest-mock==3.5.1
```

用以下指令進行測試與覆蓋率計算：

```
pipenv run pytest --cov-report term-missing --cov-config=.coveragerc
--cov=./api/ tests/
```

測試結果如下圖，這裡很容易就達到 100% 覆蓋率，因為相對爬蟲來說，API 只是從資料庫撈資料出來而已，比較少做邏輯處理，因此測試很容易撰寫：

```
sam@DESKTOP-IKT69L5:~/FinMindBook/DataEngineering/Chapter8/8.1.5$ pipenv run pytest --cov-report term-missing
--cov-config=.coveragerc --cov=./api/ tests/
Loading .env environment variables…
======================================= test session starts =======================================
platform linux -- Python 3.6.9, pytest-6.2.4, py-1.10.0, pluggy-0.13.1
rootdir: /home/sam/FinMindBook/DataEngineering/Chapter8/8.1.5, configfile: pytest.ini
plugins: cov-2.11.1, mock-3.5.1
collected 5 items

tests/test_main.py .....                                                                    [100%]

======================================== warnings summary =========================================
tests/test_main.py::test_taiwan_stock_price
  /home/sam/.local/share/virtualenvs/8.1.5-C9_5Yjkg/lib/python3.6/site-packages/pandas/io/sql.py:1431: SADepre
cationWarning: The Connection.run_callable() method is deprecated and will be removed in a future release.  In
voke the callable function directly, passing the Connection. (deprecated since: 1.4)
    self.connectable.dialect.has_table, name, schema or self.meta.schema

-- Docs: https://docs.pytest.org/en/stable/warnings.html

---------- coverage: platform linux, python 3.6.9-final-0 -----------
Name              Stmts   Miss  Cover   Missing
-------------------------------------------------
api/__init__.py       0      0   100%
api/config.py         6      0   100%
api/main.py          19      0   100%
-------------------------------------------------
TOTAL                25      0   100%

======================================= 5 passed, 1 warning in 2.65s =======================================
sam@DESKTOP-IKT69L5:~/FinMindBook/DataEngineering/Chapter8/8.1.5$
```

以下將展示測試範例：

https://github.com/FinMind/FinMindBook/blob/master/DataEngineering/
Chapter8/8.1.5/tests/test_main.py

test_main.py，關於每個測試的情境、細節，本書都寫在註解上：

```python
import time
from multiprocessing import Process

import pytest
import requests
import uvicorn
from fastapi.testclient import (
    TestClient,
)
from sqlalchemy import engine

from api.main import (
    app,
    get_mysql_financialdata_conn,
)

client = TestClient(app)
# 使用 fastapi 官方教學
# https://fastapi.tiangolo.com/tutorial/testing/
# 測試框架

# 測試對資料庫的連線,
# assert 回傳的物件, 是一個 sqlalchemy 的 connect 物件
def test_get_mysql_financialdata_conn():
    conn = (
        get_mysql_financialdata_conn()
    )
    assert isinstance(
        conn, engine.Connection
```

```
    )


# 測試對 'http://127.0.0.1:5000/' 頁面發送 request,
# 得到的回應 response 的狀態 status_code, json data
def test_read_root():
    response = client.get("/")
    assert response.status_code == 200
    assert response.json() == {
        "Hello": "World"
    }



# 測試對 'http://127.0.0.1:5000/taiwan_stock_price' 頁面發送 request,
# 並帶 stock_id, start_date, end_date 參數
# 得到的回應 response 的狀態 status_code, json data
def test_taiwan_stock_price():
    response = client.get(
        "/taiwan_stock_price",
        params=dict(
            stock_id="2330",
            start_date="2021-04-01",
            end_date="2021-04-01",
        ),
    )
    assert response.status_code == 200
    # 以下特別重要, 需要把 response 結果寫死
    # 避免未來 api schema 改變時, 影響 api 的使用者
    # 例如 Open 目前 response 是 float, 598.0,
    # 未來就不能改成 int, 598, 這是不被允許的
    # 因為 float -> int, 會大大影響使用者進行資料處理
    # 同理, int -> float 也是禁止的,
    # 資料型態在一開始就必須決定好
    assert response.json() == {
```

```
        "data": [
            {
                "StockID": "2330",
                "TradeVolume": 45972766,
                "Transaction": 48170,
                "TradeValue": 27520742963,
                "Open": 598.0,
                "Max": 602.0,
                "Min": 594.0,
                "Close": 602.0,
                "Change": 15.0,
                "Date": "2021-04-01",
            }
        ]
    }



# end to end 測試, 模擬真實使用 request 套件發送請求
# 使用 Process 開另一個進程, 模擬啟動 api
# 之後會在主進程, 對此 api 發送 request
@pytest.fixture(scope="module")
def setUp():
    proc = Process(
        target=uvicorn.run,
        args=(app,),
        kwargs={
            "host": "127.0.0.1",
            "port": 5000,
            "log_level": "info",
        },
        daemon=True,
    )
```

```
    proc.start()
    time.sleep(1)
    return 1


# 測試對 api 發送 requests,
# assert 回傳結果是 {"Hello": "World"}
def test_index(setUp):
    response = requests.get(
        "http://127.0.0.1:5000"
    )
    assert response.json() == {
        "Hello": "World"
    }
# 測試對 api 發送 requests,
# assert 回傳的 data, 這裡把真實的 case 寫下來
# 跟 test_taiwan_stock_price 概念一樣,
# 唯一的差別是, 這裡是真實場景, 對 api 發送 requests
def test_TaiwanStockPriceID(setUp):
    payload = {
        "stock_id": "2330",
        "start_date": "2021-04-01",
        "end_date": "2021-04-01",
    }
    res = requests.get(
        "http://127.0.0.1:5000/taiwan_stock_price",
        params=payload,
    )
    resp = res.json()["data"]
    assert resp == [
        {
            "StockID": "2330",
```

```
        "TradeVolume": 45972766,
        "Transaction": 48170,
        "TradeValue": 27520742963,
        "Open": 598.0,
        "Max": 602.0,
        "Min": 594.0,
        "Close": 602.0,
        "Change": 15.0,
        "Date": "2021-04-01",
    }
]
```

這裡除了使用 FastAPI 推薦的測試框架外，也同時測試，真實架設 API
Server，並發送 Request 的情境，因此使用 Process 額外開一個進程，讓
讀者能夠在同一個測試中，同時架 Server 跟發送 Request。

跟 8.1.4 一樣，也可以用 VS Code 提供的測試插鍵來做單一測試。

以上測試都是手動進行，但人都會健忘，可能在某次新版本上，忘記跑測試，那怎麼辦呢？或是另一個情境，在多人開發上，可能某些人沒跑測試就 Merge，這也是一個問題，這是工作上真實的案例，如果將以上測試，加入 GIT 中，在每次發送 Pull Request or Merge Request 時，自動跑測試，這是不是使整個開發流程更順暢呢？

下個章節，將介紹 Continuous Integration（CI），持續性整合。

CICD 持續性整合、部屬

9.1 什麼是 CICD ？

CICD 原文是，Continuous Integration、Continuous Delivery，簡單來說，就是**自動化**程式開發流程。持續將新程式、功能，**整合**進原先的專案中，這就是 CI，而 CD 就是持續將 CI 整合過後的新版本，部屬到正式（產品）環境上。

實務上的情境，團隊成員開發新功能後，經過 CICD，自動測試並部屬新版本到正式環境中，讓新功能交付這件事，有個 SOP 流程，且自動執行。

9.2　CI 持續性整合

可能有些人會認為，CI 就是單純跑測試，實際上 CI 是，持續性整合，在多人開發上，持續將團隊開發的程式、新功能，整合到主產品上，且將流程自動化。新的功能越快整合到產品上，整個開發流程會更加快速。

某些工程師可能會提出疑問，CI 不是 DevOps 的工作嗎？為什麼一般工程師需要了解？我只要專心開發新功能、新的分析結果就好了。筆者認為，工程師必須要有專業，要對自己開發的程式負責，有獨立作業的能力，不能什麼都依靠別人，而最好的負責方式就是，確保自己的程式不要發生 Error，特別是做產品、服務，來自 User 的壓力是很可怕的。

本書的主軸是，以 Data 為起點做一個產品。試想一下，如果離開公司，沒有 QA、沒有 DevOps 、沒有 SRE，你私底下開發 Side Project 、產品，有辦法保證穩定性嗎？

CI 可以協助軟體開發的流程，而 CI 工具有很多種，以下將介紹 Gitlab-CI，並以 8.1.4 的爬蟲專案為例。

9.3　Gitlab-CI、以爬蟲專案為例

這裡使用 Gitlab-CI，無法在 GitHub 上進行，必須使用 Gitlab，因此專案改到以下連結進行。

https://gitlab.com/FinMindBook/financialdata

首先，需要 **.gitlab-ci.yml** 設定檔，如下，設定檔每個步驟，都寫在註解中，這裡要注意一點，CI 測試會在 Docker 中運行，確保隔絕一切外在因素，讓測試與實際運作在同一個 Docker 環境底下：

https://gitlab.com/FinMindBook/financialdata/-/blob/9.3/.gitlab-ci.yml

```
.gitlab-ci.yml

stages:
  # CI pipeline
  # 一般開發流程
  # test 測試過了 ->
  #   merge 進 stage/master ->
  #   build 建立 docker image ->
  #   deploy 部屬服務
  # 因此 pipeline 設定順序如下
  # test -> build -> deploy
  - test
  - build
  - deploy


# CI 名稱
test-crawler:
    # 在 stage 設定 CI pipeline 順序
    # 這裡 pipeline 步驟是 test
    stage: test
    # 運行測試的 docker image
    # 由於 Dockerfile 基於
    # continuumio/miniconda3:4.3.27 下去做設定
    # 因此這裡測試也使用同樣的 image
    image: continuumio/miniconda3:4.3.27
    before_script:
      # 測試事前準備的指令
```

```
    # 需要先安裝環境
    - pip install pipenv
    - pipenv sync
    - python3 genenv.py
  script:
    # 實際測試指令
    - pipenv run pytest --cov-report term-missing --cov-config
=.coveragerc --cov=./financialdata/ tests/
  only:
      # 設定只有在 merge requests 的情況下
      # 才會運作此 CI
    refs:
      - merge_requests
```

- stage：設定 CICD pipeline 步驟，test。
- image：設定與 Dockerfile 相同的 image。
- before_script：事前指令，這邊進行環境安裝。
- script：主要指令，在這進行測試。
- only：設定在什麼情況下，才會執行該 pipeline，這邊設定在 merge_request 的情況下跑測試。

以上展示 CI Test 測試步驟，而整個 CICD 流程還包含，Build，建立 Docker Image，Deploy，部屬新版本，這在後續章節會一一介紹。

那 Test 實際上怎麼運作呢？當開發一個新功能或是對於程式做改動，對於原本的程式，都有風險，可能會導致原本的程式出錯，而測試就是避免這個問題，因此設定在 Merge 進 master ／ stage 前（refs: merge_request），會自動進行測試，不用再手動。

實際情境如下圖：

https://gitlab.com/FinMindBook/financialdata/-/merge_requests/4

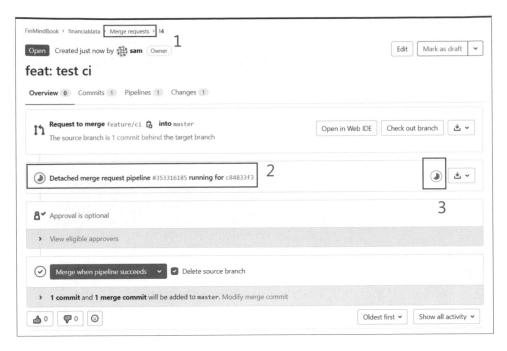

上圖 1，開發完新 Feature 後，建立 Merge Requests，在圖 2 的地方，就會自動進行 CI 測試，而測試執行的步驟，就是根據 .gitlab-ci.yml 的設定（script）去運行，可以點擊圖 3，進去觀察運作是否正常（點擊圖 3 後，會跳轉到以下頁面，需再點擊 test-crawler）。

或是直接點以下連結：

https://gitlab.com/FinMindBook/financialdata/-/jobs/1504132413

由於 GitLab-CI 背景是黑底，不適合放到書上，因此複製運行指令，並
一一講解（中間安裝環境的訊息將會省略不貼上）

```
Running with gitlab-runner 14.1.0-rc1 (e9489c8a)
  on docker-auto-scale fa6cab46
Resolving secrets
00:00
Preparing the "docker+machine" executor
00:25
Using Docker executor with image continuumio/miniconda3:4.3.27 ...
Pulling docker image continuumio/miniconda3:4.3.27 …
Using docker image sha256:2c0b9b979d1b1d487668d703310479afb6c36f1187
e832c11b775b9e3a03685e for continuumio/miniconda3:4.3.27 with digest
continuumio/miniconda3@sha256:5a2ec668104a2f465b92f89141bf8259cfc11f8972
b47c12fb70cc573620f243 ...
```

GitLab-CI 是在 Docker 環境中進行，需要 pull continuumio/miniconda3:4.3.27，
這 image 是在 .gitlab-ci.yml 中設定的。

```
$ pip install pipenv
Collecting pipenv
```

在 .gitlab-ci.yml 中，before_script，寫明需要先安裝 pipenv，因此 CI 正
在按照設定的指令運作。

```
$ pipenv sync
Creating a virtualenv for this project...
```

同樣，before_script 第二個步驟，使用 pipenv sync 安裝所需 Package。

```
$ python3 genenv.py
```

before_script 的第三個步驟，使用 genenv.py 建立環境變數 .env。

```
$ pipenv run pytest --cov-report term-missing --cov-config=.coveragerc
--cov=./financialdata/ tests/
```

此步驟，是 .gitlab-ci.yml 中 script 的指令，也就是前面章節使用的測試指令。最後運作結果如下：

```
$ pipenv run pytest --cov-report term-missing --cov-config=.coveragerc
--cov=./financialdata/ tests/
Loading .env environment variables...
========================= test session starts =========================
platform linux -- Python 3.6.2, pytest-6.2.4, py-1.10.0, pluggy-0.13.1
rootdir: /builds/FinMindBook/financialdata, configfile: pytest.ini
plugins: cov-2.11.1, mock-3.5.1
collected 18 items
tests/crawler/test_taiwan_stock_price.py ..................
[100%]
----------- coverage: platform linux, python 3.6.2-final-0 -----------
Name                                          Stmts   Miss  Cover Missing
-----------------------------------------------------------------------
financialdata/__init__.py                         0      0   100%
financialdata/backend/__init__.py                 0      0   100%
financialdata/backend/db/__init__.py              5      5     0%  1-8
financialdata/backend/db/clients.py               7      7     0%  1-18
financialdata/backend/db/db.py                    9      9     0%  1-19
financialdata/backend/db/router.py               30     30     0%  1-55
financialdata/config.py                          10     10     0%  1-13
financialdata/crawler/__init__.py                 0      0   100%
financialdata/crawler/taiwan_futures_daily.py    51     51     0%  1-141
financialdata/crawler/taiwan_stock_price.py      96      0   100%
```

```
financialdata/producer.py                              15      15     0%    1-30
financialdata/scheduler.py                             17      17     0%    1-34
financialdata/schema/__init__.py                        0       0   100%
financialdata/schema/dataset.py                        34       0   100%
financialdata/tasks/__init__.py                         0       0   100%
financialdata/tasks/task.py                             8       8     0%    1-19
financialdata/tasks/worker.py                           4       4     0%    1-13
---------------------------------------------------------------------------------
TOTAL                                                 286     156    45%
===================== 18 passed in 71.82s (0:01:11) =======================
```

跟 8.1.4 的結果一致，唯一的差別是，這是在 GitLab-CI 自動運作，不需要每次都手動。流程自動化，就是本書的目標，這可以加快開發速度與穩定性。

接下來，將介紹 Build 的步驟，建立 Docker Image，並 Push 推到 Docker Hub。

9.4　Gitlab-CI，建立 Docker Image

9.4.1　Gitlab-Runner

根據 Gitlab 官方文件：

https://docs.gitlab.com/12.10/ee/ci/docker/using_docker_build.html

在 CICD 階段，要建立 Docker Image 時，需要先設定 GitLab-Runner，那 GitLab-Runner 是什麼呢？

簡單來說，是幫你運作 CI 指令的機器，工人（Worker）的概念，以 9.3 的例子來說，Merge Request 會觸發 CI，這時 GitLab 就會發送任務給 GitLab Runner，讓 Runner 去進行測試的 Job，這跟 5.5 介紹的 Celery，是很類似的概念，都是發送 Job 給 Worker 去執行，以下將講解如何設定 GitLab Runner。

以 financialdata 的 GitLab 專案為例：

https://gitlab.com/FinMindBook/financialdata

到 Settings 點選 CI/CD：

再到 Runners 的地方點選 Expand：

就會看到 runners 的設定，如下：

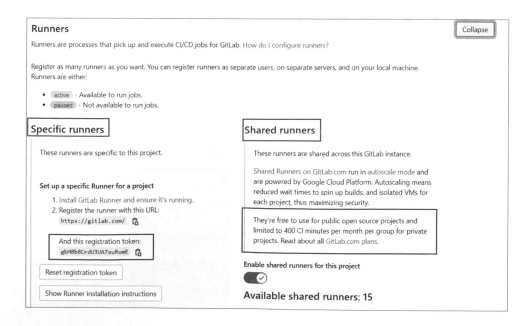

這裡有兩種 runner，**Specific runners** 和 **Shared runners**。

Specific runners 是讓使用者自行設定在特定機器上跑的 Runner，GitLab 官方也有提供教學。

Shared runners 是 GitLab 官方提供的免費 Runner，私人專案每個月有 400 分鐘的運行時間。

根 據 官 方 教 學，https://docs.gitlab.com/12.10/ee/ci/docker/using_docker_build.html#use-shell-executor，

要使用 **Specific runners**，並設定 shell runner 在指定的機器上，首先，需要安裝 Gitlab-Runner，指令如下。

```
sudo curl -L --output /usr/local/bin/gitlab-runner https://gitlab-runner-
downloads.s3.amazonaws.com/latest/binaries/gitlab-runner-linux-amd64

sudo chmod +x /usr/local/bin/gitlab-runner

sudo useradd --comment 'GitLab Runner' --create-home gitlab-runner
--shell /bin/bash

sudo gitlab-runner install --user=gitlab-runner --working-directory=/
home/gitlab-runner

sudo gitlab-runner start
```

將 Gitlab-Runner 加入 Docker 權限，以便操作 Docker：

```
sudo usermod -aG docker gitlab-runner
```

安裝完後，執行以下指令，註冊 Gitlab-Runner。

```
sudo gitlab-runner register --non-interactive --url
"https://gitlab.com/" --registration-token "gbH8b8CrdU3UA7xuRumE"
--executor "shell" --description "build_image" --tag-list "build_image"
```

執行畫面如下：

```
sam@localhost:~$ sudo gitlab-runner register --non-interactive --url "https://gitlab.com/" --registration-token "gbH8b8CrdU3UA7xuRumE" --executor "shell" --descri
ption "build_image" --tag-list "build_image"
Running in system-mode.

Registering runner... succeeded                 runner=gbH8b8Cr
Runner registered successfully. Feel free to start it, but if it's running already the config should be automatically reloaded!
sam@localhost:~$
```

最後顯示 succeeded，代表成功註冊，這時重新整理畫面：

https://gitlab.com/FinMindBook/financialdata/-/settings/ci_cd

就會看到，Specific runners 底下，多了一個 runner，這就是以上指令設定的 runner。

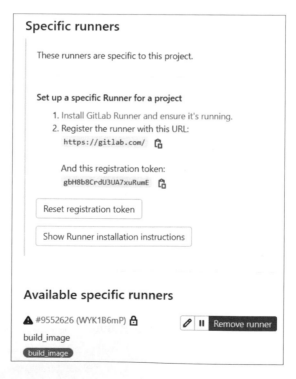

這時 runner 會先顯示驚嘆號，代表正在建立 runner，並與 GitLab 做連線，等個約 3~5 分鐘，就會成功了，如下

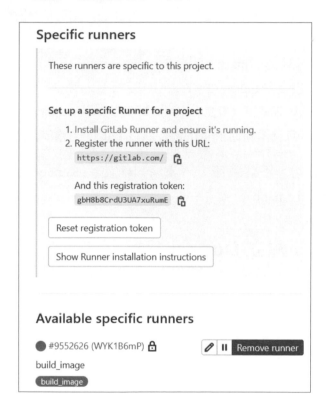

註冊指令蠻長的，本書一一講解。

- register：顧名思義，就是註冊 GitLab-Runner。
- --non-interactive：非互動模式，互動模式指的是，在註冊 GitLab-Runner 的階段，會被詢問關於設定細節等多個問題，如果加入此指定，就可以像一鍵執行，中間不會再詢問（不會再互動）多個問題。
- --url：顧名思義，就是註冊的 url，這裡選 GitLab 的官方網址。
- --registration-token：**Specific runners** 底下提供的 token。

- --executor：指定在什麼環境下執行，這裡選擇 shell，也就是一般 Linux 原生環境下，根據官方文件，https://docs.gitlab.com/runner/executors/，還有 Docker、Kubemetes 等選擇，其中 Docker 在接下來的章節會用到。
- --description：關於此 Runner 的描述，像書中設定 build_image，表示這 Runner 是給建立 Image 使用的。
- --tag-list：關於此 Runner 的 Tag，一般情境，會進行 Unit Test、Build Image、Deploy 等等的 CICD，因此會建立多個 Runner，那在 .gitlab-ci.yml 中，要如何選擇特定的 Runner 執行 CICD ？這時就是依靠 Tag。

建立好 Runner 後，下個章節，將講解如何使用 Runner，並在 CICD 階段，建立 Docker Image。

9.4.2　CICD 建立 Docker Image

在 7.5，書中介紹了 Docker Swarm，在此架構下，爬蟲、API 都是用 Docker 包裝好，部屬上線，如果要發布新版本，都是藉由 Docker 進行。所以在第二步驟，需要 Build Docker Image，才能發布新版本，以下將介紹如何在 CICD 底下，將 Build Image 這件事，達到自動化。

與 9.3 相同，在 .gitlab-ci.yml 寫入，建立 Docker Image 的設定，如下

https://gitlab.com/FinMindBook/financialdata/-/blob/9.4.2/.gitlab-ci.yml

.gitlab-ci.yml
```
stages:
  # CI pipeline
  # 一般開發流程
  # test 測試過了 ->
  #   merge 進 stage/master ->
  #   build 建立 docker image ->
```

```
#   deploy 部屬服務
# 會根據此順序依序進行
# test -> build -> deploy
- test
- build
- deploy


# test-crawler 的部分略過


# CI 名稱, 建立 docker image
build-docker-image:
    # 在 stage 設定 CI pipeline 順序
    # 這裡 pipeline 步驟是 build
    stage: build
    before_script:
        # 由於 build 好 image 後，需要 push 到 docker hub
        # 需要事先登入
        # 登入部分，這裡使用 token 做登入，可參考以下連結
        # https://docs.docker.com/docker-hub/access-tokens/#create-an-
access-token
        # token 你可以使用明碼，或是存入 gitlab CICD Variables，用變數的方式引用
        # 如下
        # docker login -u "linsamtw" -p 7777-7777-7777-7777
        # 但實務上，不希望 token 公開，因此會採用以下方式
        - docker login -u "linsamtw" -p ${DOCKER_HUB_TOKEN}
    script:
        # 建立 image
        - docker build -f Dockerfile -t linsamtw/crawler:9.4 .
        # push
        - docker push linsamtw/crawler:9.4
    tags:
        # 設定使用前面建立的 runner 執行
```

```
    - build_image
  only:
    # 設定只有在 staging, master 階段
    # 才會 build image
    refs:
      - staging
      - master
```

- stage：設定 CICD pipeline 步驟，build。
- image：不設定，因為這裡要使用 9.4.1 註冊的 GitLab-Runner。
- before_script：事前指令，因為需要 Push Docker Image，先做登入動作。
- script：主要指令，在這建立 Image 與 Push 到 Docker Hub。
- tag：設定要用哪個 GitLab-Runner 執行該段 Pipeline。
- only：設定在 staging、master 階段，執行 build pipeline

這裡有一點需特別注意，由於需要 Push Docker Image 到 Docker Hub，需事先登入，這裡使用 Token 做登入，如下

```
docker login -u "linsamtw" -p ${DOCKER_HUB_TOKEN}
```

實務上，為了安全性，關於密碼、金鑰、Token 等資訊，都會存到 CICD Variables 中，以下先介紹如何拿到 Docker Token，接著再講解如何在 GitLab 設定 CICD Variables。

Docker Token 申請方式，以下是官方教學，https://docs.docker.com/docker-hub/access-tokens/#create-an-access-token，簡單來說，到以下連結 https://hub.docker.com/settings/security

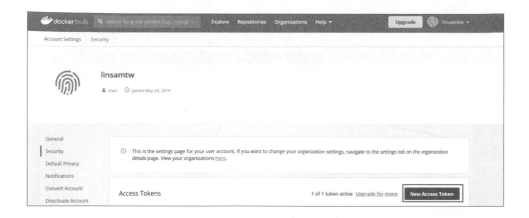

點選 New Access Token，就會看到以下畫面，接著建立你自己 Token

接著將 Token 存起來，到以下畫面，設定 Variables：

點選 Add variable：

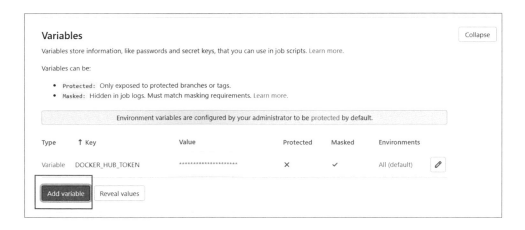

接著到 Key 的地方，輸入你想要的 Variable 名稱，這裡輸入 DOCKER_ HUB_TOKEN，接著在 Value 的地方，輸入你的 Token。

在 Flags 地方，有兩個選項 Protect variable、Mask variable。

- Protect variable 用途是，限定此 CICD Variables 只能在受保護的 branches 中使用，這裡先不勾選。
- Mask variable 用途是，會在 CICD 過程，將 Variables 遮住，避免 Token 外洩。

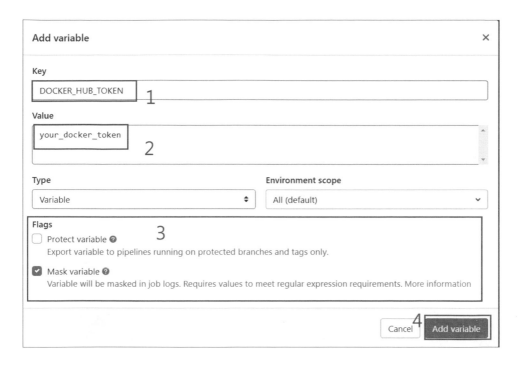

當然如果不想這麼複雜，可以先將 Token 直接使用明碼，但比較好的做法，還是用 CICD Variable 做管理。

接著當 Merge Request 合併到 master 後，讀者可能會想問，要去哪裡查看 Build Image 步驟？成功或是失敗的 Log ？

如下圖，先到 CICD，選擇 Jobs，可以看到 Name 為 build-docker-image，
這裡的 Name，就是 .gitlab-ci.yml 中設定的名稱，接著點選圖 4，
passed，就可以看到實際上，Build 執行的指令。

確認成功後，到 Docker Hub，就可以看到，CICD 上傳的 Docker Image。

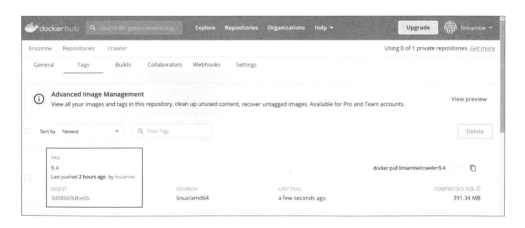

這時完成了 test、build 兩個步驟，且都達到自動化，不需要再手動跑測
試、手動建立 Docker Image，下個章節，將展現第三步驟，部屬新版
本，將 deploy 這步驟，也進行自動化。

9.5 Gitlab-CI，部屬新版本

首先，回到 7.5，本書建立了 Docker Swarm、並用 Portainer 作為 UI 介面，http://139.162.104.54:9000/#/services

同樣以 financialdata 為例，financialdata_cralwer_twse、financialdata_scheduler，目前 Image 版本停留在，linsamtw/crawler:7.2.1，本章節將使用 CICD，協助 Deploy 部署到 Swarm 中，達到自動化。

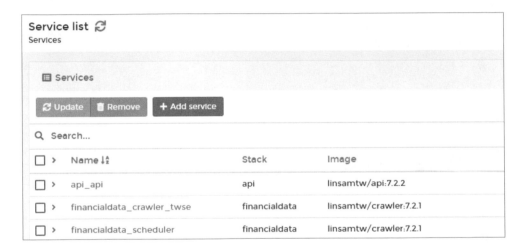

先展示 CICD 自動 Deploy 的成果，如下圖，以下 financialdata_crawler_twse，Image 從 7.2.1 更新版本到 9.5，那實際上怎麼運作的呢？

同樣，在 .gitlab.ci.yml 中，加入第三步驟，Deploy。

https://gitlab.com/FinMindBook/financialdata/-/blob/9.5/.gitlab-ci.yml

.gitlab-ci.yml

```
stages:
  # CI pipeline
  # 一般開發流程
  # test 測試過了 ->
  #   merge 進 stage/master ->
  #   build 建立 docker image ->
  #   deploy 部屬服務
  # 會根據此順序依序進行
  # test -> build -> deploy
  - test
  - build
  - deploy


# 一樣略過 test-crawler


# CI 名稱, 建立 docker image
build-docker-image:
    # 在 stage 設定 CI pipeline 順序
    # 這裡 pipeline 步驟是 build
    stage: build
    before_script:
      # 由於 build 好 image 後，需要 push 到 docker hub
      # 需要事先登入
      # 登入部分，這裡使用 token 做登入，可參考以下連結
      # https://docs.docker.com/docker-hub/access-tokens/#create-an-
access-token
      # token你可以使用明碼，或是存入gitlab CICD Variables，用變數的方式引用
```

```
      # 如下
      # docker login -u "linsamtw" -p 7777-7777-7777-7777
      # 但實務上,不希望 token 公開,因此會採用以下方式
      - docker login -u "linsamtw" -p ${DOCKER_HUB_TOKEN}
    script:
      # 建立 image
      - make build-image
      # push
      - make push-image
    tags:
      # 設定使用前面建立的 runner 執行
      - build_image
    only:
      # 設定只有在下 tag 後
      # 才會 build image
      - tags


# CI 名稱, deploy crawler
deploy-crawler:
    # 在 stage 設定 CI pipeline 順序
    # 這裡 pipeline 步驟是 deply
    stage: deploy
    script:
      # 部屬爬蟲
      - make deploy-crawler
    tags:
      # 設定使用前面建立的 runner 執行
      - build_image
    only:
      # 設定只有在下 tag 後
      # 才會 deploy
      - tags
```

本章節做了兩個更動：

1. only 部分，從 stage ／ master 改成 tags，跟發布／部屬新版本有關，在這種情境下，Merge 進 master 不會部屬，需要下 Tag 版本號，才會開始建立 Image，並部屬上線。主要是情境不同。

2. script 部分，改用 make 指令，主要是因為，每次 build image 的版本號會不同，在 9.4.2 章節，是直接寫死版本號，但在開發過程中，會持續建立新版本的 Docker Image，因此改用 Make，並在 Makefile 加入，版號更動這件事，如下。

這裡只列出本章節使用到的指令，其餘可直接參考連結

https://gitlab.com/FinMindBook/financialdata/-/blob/9.5/Makefile

```Makefile
GIT_TAG := $(shell git describe --abbrev=0 --tags)


# 建立 docker image
build-image:
    docker build -f Dockerfile -t linsamtw/crawler:${GIT_TAG} .


# 推送 image
push-image:
    docker push linsamtw/crawler:${GIT_TAG}


# 部屬爬蟲
deploy-crawler:
    GIT_TAG=${GIT_TAG} docker stack deploy --with-registry-auth -c
crawler.yml financialdata


# 部屬 scheduler
deploy-scheduler:
```

```
    GIT_TAG=${GIT_TAG} docker stack deploy --with-registry-auth -c
scheduler.yml financialdata
```

- GIT_TAG：在這拿到 GitLab 的 Tag 編號，作為後續 Docker Image 的版本號。
- build-image：加入 ${GIT_TAG}，這樣才能引用 GIT_TAG 變數，根據此建立不同版號的 Image。
- push-image：一樣加入 ${GIT_TAG}，根據 GIT_TAG 推送不同版號的 Image。
- deploy-crawler：加入 GIT_TAB=${GIT_TAG}，將 GIT_TAG 傳入 crawler.yml。
- deploy-scheduler：同上。

crawler.yml 在 Image 的部分，同樣也做了更動，加入 GIT_TAG，作為版號控制，並在 Makefile 的指令中，將 GIT_TAG 傳進去，如下。

https://gitlab.com/FinMindBook/financialdata/-/blob/9.5/crawler.yml

```
crawler.yml
version: '3.0'
services:
  crawler_twse:
    image: linsamtw/crawler:${GIT_TAG}
    hostname: "twse"
    command: pipenv run celery -A financialdata.tasks.worker worker
--loglevel=info --concurrency=1  --hostname=%h -Q twse
    restart: always
    # swarm 設定
    deploy:
      mode: replicated
      replicas: 1
```

```
    placement:
      constraints: [node.labels.crawler_twse == true]
   environment:
    - TZ=Asia/Taipei
   networks:
      - my_network


networks:
  my_network:
    # 加入已經存在的網路
    external: true
```

完成以上更動並 Merge 到 master 後，新增一個 Tag，CICD 就會自動執行 build 與 deploy 步驟，如下：

https://gitlab.com/FinMindBook/financialdata/-/tags

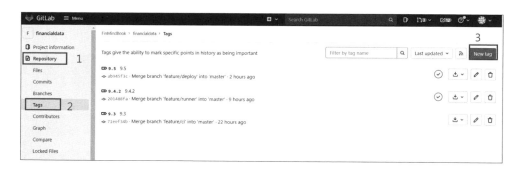

新增好 Tag 後，再到 CI/CD 中的 Piplines，就能看到執行的步驟了，第一個是 build。

第二個是 deploy。

這時切換到 Portainer 頁面，就能看到，financialdata_crawler_twse 已經換成新的 Image 了。

financialdata_scheduler 部分，讀者可自行仿造，或是直接參考以下連結：

https://gitlab.com/FinMindBook/financialdata/-/blob/9.5.2/.gitlab-ci.yml

成功更新畫面如下：

	Name ↕	Stack	Image	Scheduling Mode	Published Ports
☐ ﹥	api_api	api	linsamtw/api:7.2.2	replicated 1 / 1 ↕ Scale	☑ 8888:8888
☐ ﹥	financialdata_crawler_twse	financialdata	linsamtw/crawler:9.5.2	replicated 1 / 1 ↕ Scale	-
☐ ﹥	financialdata_scheduler	financialdata	linsamtw/crawler:9.5.2	replicated 1 / 1 ↕ Scale	-

到此步驟，將 test - 測試、build - 建立 Image、deploy - 部屬上線，都完成自動化了，即使多人開發，也只需要按個按鈕，建立新的 Tag，其他成員也能自行換版，且保證通過測試，完成了 CICD 自動化流程。

下一章節，9.6，本書繼續介紹 API 專案的 CICD 流程。

9.6　Gitlab-CI、以 API 專案為例

9.6.1　Test - 測試

API 測試，包含對資料庫連線的 End to End 測試，但資料庫，一般都會設防火牆，只限特定 IP 做連線，必須使用 Gitlab-Runner 來執行測試，無法使用免費的 Share Runner。

上一章節，註冊了 shell 類型的 Gitlab Runner，本章節，需要執行測試，要建立 Docker 類型的 Gitlab Runner。

先到 api 專案底下的 CICD - runner 底下，獲取該專案的 Token。

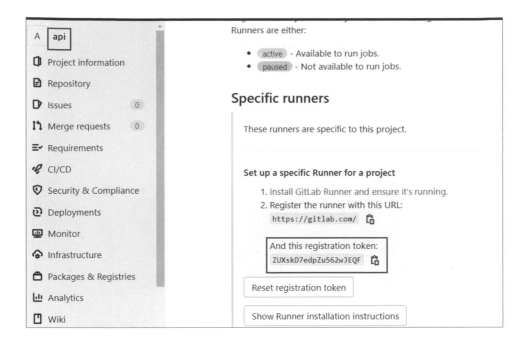

再執行以下指令：

```
sudo gitlab-runner register --non-interactive --url "https://gitlab.
com/" --registration-token "ZUXskD7edpZu562wJEQF" --executor "docker"
--docker-image continuumio/miniconda3:4.3.27 --description "docker-
runner" --tag-list "docker-runner"
```

這裡有幾點與註冊 shell runner 不同。

- --executor：指定在 Docker 環境下執行。
- --docker-image：指定 Docker 的 image。

執行後，等個 5 分鐘，看到以下畫面，就代表建立成功了。

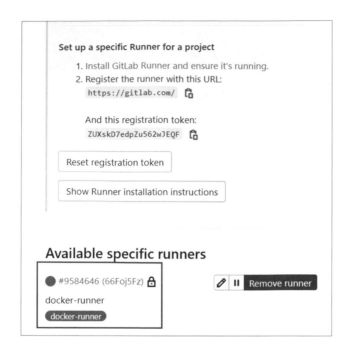

有了專屬的 Runner 後，就可以針對資料庫設定，只對這個 Runner 所在的機器，開設防火牆，這樣可以同時達到，安全性與執行測試的條件了。

關於 .gitlab-ci.yml，寫法與 9.5 類似，如下，或可參考以下連結。

https://gitlab.com/FinMindBook/api/-/blob/9.6.1/.gitlab-ci.yml

```
stages:
  # CI pipeline
  # 一般開發流程
  # test 測試過了 ->
  #   merge 進 stage/master ->
  #   build 建立 docker image ->
  #   deploy 部屬服務
  # 會根據此順序依序進行
  # test -> build -> deploy
```

```
    - test
    - build
    - deploy
# CI 名稱，test 步驟
test-api:
    # 在 stage 設定 CI pipeline 順序
    # 這裡 pipeline 步驟是 test
    stage: test
    # 運行測試的 docker image
    # 由於 Dockerfile 基於
    # continuumio/miniconda3:4.3.27 下去做設定
    # 因此這裡測試也使用同樣的 image
    image: continuumio/miniconda3:4.3.27
    before_script:
        # 測試事前準備的指令
        # 需要先安裝環境
        - pip install pipenv
        - pipenv sync
        - python3 genenv.py
    script:
        # 實際測試指令
        pipenv run pytest --cov-report term-missing --cov-config
=.coveragerc --cov=./api/ tests/
    only:
        # 設定只有在 merge requests 的情況下
        # 才會運作此 CI
        refs:
            - merge_requests
```

基本上指令都類似，Test 執行結果，也可到以下畫面查看：

https://gitlab.com/FinMindBook/api/-/merge_requests/1

9-31

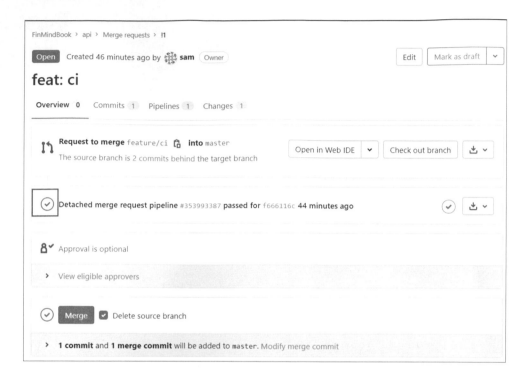

到這步驟，CI 自動化測試，已經完成，即使資料庫有設防火牆，只要對該機器上的 Runner 開權限，一樣能執行 CI。下一章節，將介紹自動化建立 Docker Image。

9.6.2 Build - 建立 Image

在這步驟，一樣要先建立 shell GitLab Runner，因為在 9.4 建立的，是專門給 financialdata 專案使用，無法給 api 專案使用，因此使用同樣指令，來建立 api 專屬的 shell GitLab Runner。

```
sudo gitlab-runner register --non-interactive --url "https://gitlab.
com/" --registration-token "ZUXskD7edpZu562wJEQF" --executor "shell"
--description "build_image" --tag-list "build_image"
```

CI 設定檔 .gitlab-ci.yml 如下：

https://gitlab.com/FinMindBook/api/-/blob/9.6.2/.gitlab-ci.yml

```
stages:
  # CI pipeline
  # 一般開發流程
  # test 測試過了 ->
  #   merge 進 stage/master ->
  #   build 建立 docker image ->
  #   deploy 部屬服務
  # 會根據此順序依序進行
  # test -> build -> deploy
  - test
  - build
  - deploy


# test-api 略過


# CI 名稱, 建立 docker image
build-docker-image:
    # 在 stage 設定 CI pipeline 順序
    # 這裡 pipeline 步驟是 build
    stage: build
    before_script:
        # 由於 build 好 image 後，需要 push 到 docker hub
        # 需要事先登入
        # 登入部分，這裡使用 token 做登入，可參考以下連結
        # https://docs.docker.com/docker-hub/access-tokens/#create-an-
access-token
        # token你可以使用明碼，或是存入gitlab CICD Variables，用變數的方式引用
        # 如下
        # docker login -u "linsamtw" -p 7777-7777-7777-7777
        # 但實務上，不希望 token 公開，因此會採用以下方式
```

```
    - docker login -u "linsamtw" -p ${DOCKER_HUB_TOKEN}
  script:
    # 建立 image
    - make build-image
    # push
    - make push-image
  tags:
    # 設定使用前面建立的 runner 執行
    - build_image
  only:
    # 設定只有在下 tag 後
    # 才會 build image
    - tags
```

在下圖步驟 1，新增 tag 後，確認下圖 3，CD Build 是成功的，就完成自動化建立 Image 這步驟了。

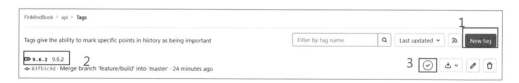

接著到你的 Docker Hub 底下，就能看到，新版本 9.6.2 的 api Docker Image，已經成功上傳了，下圖是筆者的 Docker Hub。

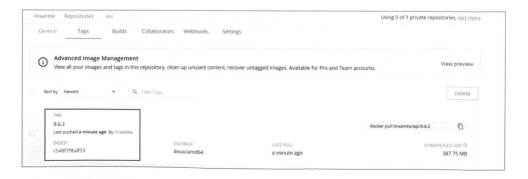

到這步驟，已經將測試、建立 image 自動化，一切都與爬蟲專案類似，
甚至連 .gitlab-ci.yml 設定檔都沒做太多更動，以上都可以複製到讀者的
專案中，讓你的專案也達到自動化。

下一章節，開始進行新版本部屬的自動化。

9.6.3 deploy - 部屬

先到 portainer，確認目前 api 版本，如下：

目前是 7.2.2 版本，等將自動化部屬做好後，只要新增 Tag，就會自動更
新版本。

設定也與 9.5 相同，如下，也是在 CICD 中使用 Make 指令執行，將 GIT_
TAG 包進 Makefile 中，根據 Tag 更動 Docker Image 版本號。

.gitlab-ci.yml

https://gitlab.com/FinMindBook/api/-/blob/9.6.3/.gitlab-ci.yml

```
stages:
  # CI pipeline
  # 一般開發流程
  # test 測試過了 ->
```

```
#   merge 進 stage/master ->
#   build 建立 docker image ->
#   deploy 部屬服務
# 會根據此順序依序進行
# test -> build -> deploy
- test
- build
- deploy


# test-api 步驟省略

# CI 名稱, 建立 docker image
build-docker-image:
    # 在 stage 設定 CI pipeline 順序
    # 這裡 pipeline 步驟是 build
    stage: build
    before_script:
        # 由於 build 好 image 後，需要 push 到 docker hub
        # 需要事先登入
        # 登入部分，這裡使用 token 做登入，可參考以下連結
        # https://docs.docker.com/docker-hub/access-tokens/#create-an-
access-token
        # token你可以使用明碼，或是存入gitlab CICD Variables，用變數的方式引用
        # 如下
        # docker login -u "linsamtw" -p 7777-7777-7777-7777
        # 但實務上，不希望 token 公開，因此會採用以下方式
        - docker login -u "linsamtw" -p ${DOCKER_HUB_TOKEN}
    script:
        # 建立 image
        - make build-image
        # push
        - make push-image
```

```
    tags:
        # 設定使用前面建立的 runner 執行
        - build_image
    only:
        # 設定只有在下 tag 後
        # 才會 build image
        - tags

# CI 名稱, deploy api
deploy-api:
    # 在 stage 設定 CI pipeline 順序
    # 這裡 pipeline 步驟是 deply
    stage: deploy
    script:
        # 部屬 api
        - make deploy-api
    tags:
        # 設定使用前面建立的 runner 執行
        - build_image
    only:
        # 設定只有在下 tag 後
        # 才會 deploy
        - tags
```

Makefile 如下：

https://gitlab.com/FinMindBook/api/-/blob/9.6.3/Makefile

```
GIT_TAG := $(shell git describe --abbrev=0 --tags)

# 建立 docker image
build-image:
    docker build -f Dockerfile -t linsamtw/api:${GIT_TAG} .
```

```
# 推送 image
push-image:
    docker push linsamtw/api:${GIT_TAG}

# 部屬 api
deploy-api:
    GIT_TAG=${GIT_TAG} docker stack deploy --with-registry-auth -c api.
yml api
```

將以上設定，加入 master 後，新增 Tag 步驟如下。

到 api 專案底下，Repository -> Tags，New tag（新增一個 Tag），這邊新增的 Tag 是 9.6.3，且等到下圖 5，CD 運作結束後，再到 Portainer 畫面查看 api 的 Docker Image 是否有更新版本。

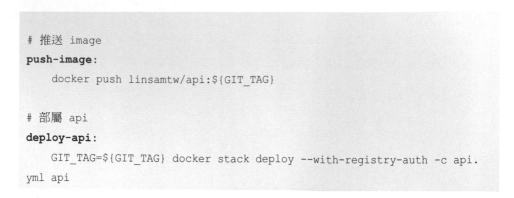

可到以下畫面，查看 CD 是否正常運作，以下為成功畫面，建立 Docker Image（build-docker-image），部屬新版本 api（deploy-api）。

以下為 Portainer 畫面，成功將 api 版本，更新到 9.6.3。

| Service list ↻ | | | | |
| Services | | | | |

☐	>	Name ↓↑	Stack	Image	Scheduling Mode	Published Ports
☐	>	api_api	api	linsamtw/api:9.6.3	replicated 1 / 1 ↕ Scale	⎘ 8888:8888

到這步驟後，關於 api 的自動化測試、建立 Docker Image、部屬新版本，都完成自動化了，讀者可以實際上體會看看，自動化的差別。單純建立新的 Tag，就會自動發布 Image 並部屬新版本，這對於後續的開發維運上，是非常方便的。

9.7 總結

第 9 章節的內容都完成後，已經達成，**爬蟲、API**，自動化測試（**CI**）、**自動化部屬（CD**），接著讀者只需要專注開發新爬蟲、新 API，剩下換版過程都定下來了，且使用分散式，因此也不用擔心效能問題。

特別的是自動化執行測試，可以確保在長期開發上專案的穩定性，不會發生團隊成員粗心大意，導致上線運作的程式，因為某次 Merge 後，掛掉無法正常運作。也避免發生團隊成員手動部屬新版本，粗心導致線上服務出問題。以上都是筆者親身經驗，因為缺少測試，導致新版本上線後，系統出現異常。

可能有讀者認為，手動也能正常運作，會出錯都是某些工程師的專業度不夠，或是刻意搞亂的。這句話是沒錯，但是，人工介入一定會有出錯的可能，而本書使用 CICD 建立了 SOP 流程，減少團隊開發上的成本，團隊只需專注在功能開發上，照著固定的流程做開發，用最有效率的方式做開發，這也是工程師的價值之一，不只是 Coding 而已，完善流程、制定架構會比單純寫程式來的重要。

第 3 篇
API 產品上線

API 服務網址

10.1 為什麼需要網址？

本章節，介紹如何設定 API 的服務網址，讓產品上線，那為什麼需要設定服務網址呢？有以下幾點好處，

第一，當需要提供其他人，讀者開發的 API 的服務時，相較於直接給對方 API 的 IP，用網址會相對好記憶，你不會希望連 google 網站，是用 142.251.42.227:443 這樣的 IP 去記憶，用以下網址跟 IP 相比，https://www.google.com.tw/，對一般人理解上，會更友善。

第二，SSL 安全性憑證。一般使用 IP 連上網時，都是 http，那要如何將 http 變成 https 呢？就需要 SSL 憑證，而這基本上需要有網址才能做設定，SSL 有什麼好處？這是現代標準的加密技術，一般進行網路瀏覽，可能會傳送一些個資，如信用卡資訊，那這中途，是有可能被攔截、擷取特定資訊，因此才需要經過加密階段。即使不考慮這些相對複雜的問題，如果不用 SSL，對於你的網站在搜尋引擎上，是會被扣分的，甚至不讓你出現在搜尋列表中。

第三，搜尋引擎。當你使用 IP 做為網站時，搜尋引擎是不會收錄的，也就是說，使用者無法透過搜尋，找到你的 API 或是網站。

實際上用網址，甚至 DNS，都有更多好處，例如 Email、子網域設定、Loading Balance 等等，就不在此多做介紹，初期只要知道，用網址好記憶、SSL 安全性、搜尋引擎相對高分，這幾點就夠了。

10.2　No-Ip 免費的網址申請

一般網址都需要花錢購買，但多數讀者沒有商用，單純在 Side Project 階段，因此盡可能使用免費服務，以下將介紹 No-IP，這個免費的網址申請網站，https://www.noip.com/。

先用以上網址註冊並登入後，到以下網址 https://my.noip.com/。

點選 **Dynamic DNS** 後，再點選 **No-IP Hostnames**。

接著點擊 **Create Hostname**，建立網址。

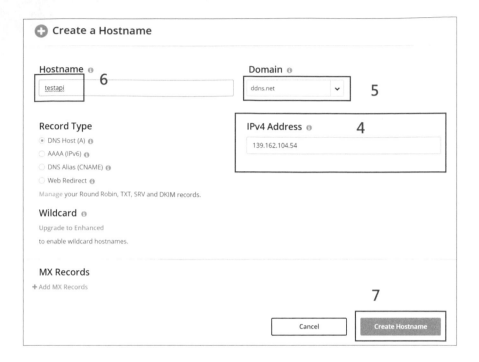

在 **IPv4 Address** 輸入你的 Linode 的 IP，**Domain** 隨便選一個你喜歡的，**Hostname** 也自訂，這裡輸入 **testapi**，最後點擊 Create Hostname。

接著就設定完成了，但這時因為 API 的 Port 設定為 8888，無法直接連上。

因為一般網站連線，預設都是連 80 Port（http）或是 443 Port（https），所以到 portainer 上，將 api 的 Port 改為 80，

先到 Services > api_api 底下，點選 Network & published ports：

將 Host port 改為 80：

改完後，點選 Apply changes：

這時再到剛剛設定的網址，http://testapi.ddns.net/docs，就能看到 API 了，是不是很簡單？

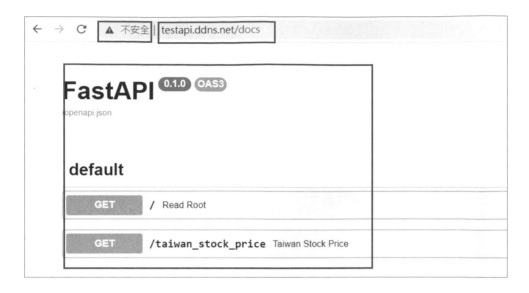

這裡注意一點，在圖上 testapi.ddns.net/docs 前面，有個三角形驚嘆號，顯示不安全，因為目前使用的是 http，下一章節，將會講解如何設定 SSL。

10.3 Let's Encrypt 免費的 SSL 憑證

同樣，SSL 也有免費與付費版本，本書介紹 Let's Encrypt，免費的 SSL 憑證，一般常見的是使用 Nginx 作設定，進階會搭配 HAProxy 做 Loading Balance，但現代出現 Swarm、K8s 等集群容器架構後，衍生出相對應的工具，本書介紹 **Traefik**，能更好的管理所有服務的網址、DNS，並自動更新 SSL，且有不錯的 Dashboard 工具，監控所有網址狀態，如下圖。

以下將介紹，如何將 API 結合 Traefik，自動完成 SSL，將你的 API 網址，加入 https。

10.4 Traefik

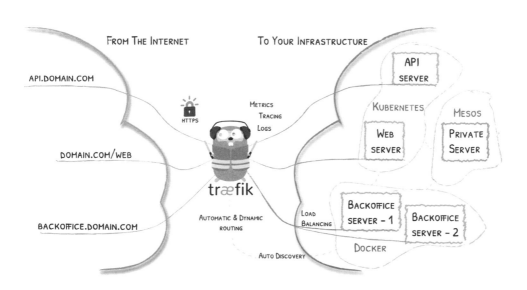

以上圖片引用 Traefik 官方網站，https://doc.traefik.io/traefik/。

中間是 Traefik 圖示，代表在 Swarm 中，Manager 的機器。Traefik 左邊是讀者設定的各種網址，想像一個情境，團隊中，開發的產品包含 API、Web、Backend，以上三個都需要相對應的網址，如 api.com、web.com、backend.com，而 Traefik 右邊，則是是相對應的服務，api_server、backend_server、web_server。

一般使用者，會藉由不同的網址，拜訪你的網站、API，如左圖，經過 Traefik 後，Traefik 會根據不同網址，回傳不同的服務。如果拜訪的網址是 api.com，Traefik 就會回應 api_server，如果拜訪的是 web.com，Traefik 就會回應 web_server。基本上是由 Traefik 統一管理，使用者全部都從同一個入口，Traefik，拜訪你的服務，這跟前面提的 Manager 概念很像，現代架構，都是講求統一管理，以下將講解如何實作。

首先，先架設 Traefik，筆者一樣使用 Docker 架設，設定檔如下：

https://github.com/FinMind/FinMindBook/blob/master/DataEngineering/Chapter10/10.4/traefik.yml

```
traefik.yml
version: "3.8"

services:

  traefik:
    image: "traefik:v2.2" # image 版本
    command:
      # 設定顯示 dashboard
      - --api.insecure=true
```

```
  - --api.dashboard=true
  # log 模式，有 DEBUG、ERROR、INFO 等模式
  - --api.debug=true
  - --log.level=ERROR
  # https://doc.traefik.io/traefik/v2.2/routing/providers/docker/
  # 根據官方文件，有 docker、k8s 等模式可做選擇
  # 在此選擇 docker
  - --providers.docker=true
  - --providers.docker.endpoint=unix:///var/run/docker.sock
  - --providers.docker.swarmMode=true
  - --providers.docker.exposedbydefault=false
  - --providers.docker.network=traefik-public
  # 進入點，80、443 分別對應 http、https
  - --entrypoints.web.address=:80
  - --entrypoints.web-secured.address=:443
  # SSL 設定
  - "--certificatesresolvers.myresolver.acme.httpchallenge=true"
  - "--certificatesresolvers.myresolver.acme.httpchallenge.
entrypoint=web"
  - "--certificatesresolvers.myresolver.acme.email=
samlin266118@gmail.com"
  - "--certificatesresolvers.myresolver.acme.storage=/letsencrypt/
acme.json"
  ports:
    # 開對外 port，除了 80、443 是給 http、https 外
    # 8080 port 給 dashboard
    # 但 8080 已經給 phpmyadmin
    # 因此這裡 published 到 8888
    - target: 80
      published: 80
      mode: host
```

```yaml
      - target: 443
        published: 443
        mode: host
      - target: 8080
        published: 8888
        mode: host
    volumes:
      # 存放 SSL 憑證
      - "letsencrypt:/letsencrypt"
      - "/var/run/docker.sock:/var/run/docker.sock:ro"
    restart: unless-stopped
    # log 最大 size
    logging:
        driver: "json-file"
        options:
            max-size: "50m"
    # deploy 設定
    deploy:
      replicas: 1
      update_config:
        parallelism: 1
        delay: 10s
        order: stop-first
        failure_action: rollback
      placement:
        max_replicas_per_node: 1
        constraints: [node.role == manager]
    networks:
        - traefik-public

networks:
  traefik-public:
```

```
    external: true

volumes:
  letsencrypt:
```

本書使用 Traefik 2.2 版本，版本 1 跟 2 存在蠻大的差異。

Log 模式選用 ERROR，如果在測試階段，可以選用 DEBUG 或是 INFO，觀察 Traefik 的 Log。Providers 選用 Docker，entrypoints 進入點設定 80 與 443，分別是 http 與 https。另外需要將 SSL 中的 Email 改成讀者自己的，而 ACME 證書存放在 /letsencrypt/acme.json，其餘就是 Swarm 的一些設定，由於前面章節都介紹過了，相信到了第十章節，讀者也很熟悉了，以下直接進行部屬。

由於 Traefik 要使用 80 Port，因為先前已將 80 Port 用於 api，在此先進行回收，如下，關閉 api 的 Published Ports：

	Name ⇅	Stack	Image	Scheduling Mode	Published Ports
☐ ›	api_api	api	linsamtw/api:10	replicated 1 / 1 ⇕ Scale	-

接著建立 Traefik 專屬的 Network：

```
docker network create --driver=overlay traefik-public
```

再部屬 Traefik：

```
docker stack deploy -c traefik.yml tr
```

部屬完成後，到 http://139.162.104.54:8888/dashboard/#/，會看到以下畫面：

Traefik 好處是有 UI 介面，讀者可以試著在上面先看看各種功能，相較於 Nginx，Traefik 與其他工具有更好的界接，如常見的監控工具 Promethus、Grafana，以上兩個服務也可以使用 Docker 進行架設，本書就不多做介紹，讀者可自行研究。

架設完 Traefik 後，API 需要做額外設定，才能跟 Traefik 結合，別擔心，本書也會一一介紹如何設定。

10.5　API 結合 Traefik

前一個章節，介紹 Traefik，統一了 User 的訪問入口，因此 API 不需要把 Port 串出去，取而代之的是，在 API 的 docker-compose.yml 做設定，讓 Traefik 認得 API，當使用者拜訪 http://testapi.ddns.net/docs（前面申請的網址），Traefik 會自動導向 API 這個 Server，設定檔如下：

https://github.com/FinMind/FinMindBook/blob/master/DataEngineering/
Chapter10/10.5/api/api.yml

```yml
api.yml
version: '3.0'
services:
  api:
    image: linsamtw/api:10
    hostname: "api"
    restart: always
    # swarm 設定
    deploy:
      mode: replicated
      replicas: 1
      placement:
        constraints: [node.labels.api == true]
      labels:
        - traefik.enable=true
        - traefik.frontend.passHostHeader-true
        # routers 設定，
        # 這裡根據 fastapi 所需的 router 做額外設定
        # fastapi 預設在 /docs 作為文件頁面，
        # 並根據 /openapi.json 去製作文件
        # 因此需額外設定
        # testapi.ddns.net/docs 與
        # testapi.ddns.net/openapi.json
        # 可被 traefik 訪問，方法如下
        # 指定 Host 與 Path
        # &&：代表 and，將 Host 與 Path 連在一起使用
        # ||：代表 or，在多個 router 時使用
        - traefik.http.routers.api-https.rule=
```

```
        Host(`testapi.ddns.net`) && PathPrefix(`/docs`) ||
        Host(`testapi.ddns.net`) && PathPrefix(`/openapi.json`) ||
        Host(`testapi.ddns.net`)
    # api loading balance 的 port
    # traefik 的好處之一，會幫你處理分流
    - traefik.http.services.api-https.loadbalancer.server.port=80
    # SSL 設定
    - traefik.http.routers.api-https.tls.certresolver=myresolver
    # http 自動導向 https
    - "traefik.http.routers.api-http-catchall.rule=hostregexp
(`{host:.+}`)"
    - "traefik.http.routers.api-http-catchall.entrypoints=web"
    - "traefik.http.routers.api-http-catchall.middlewares=api-
redirect-to-https@docker"
    - "traefik.http.middlewares.api-redirect-to-https.redirectscheme.
scheme=https"
  environment:
    - TZ=Asia/Taipei
  networks:
    - my_network
    - traefik-public

networks:
  my_network:
    # 加入已經存在的網路
    external: true
  traefik-public:
    # 需加入 traefik 的網路
    external: true
```

以上跟 8.1.5 的 api.yml 相比，移除 Port，因為改走 Traefik。多了 labels 的設定，主要都是跟 Traefik 有關，其中

traefik.enable=true：啟動 Traefik。

traefik.frontend.passHostHeader=true：傳送 Host Header 給 API Server。想像一個情境，如果 API 會記錄每一個 Request 的資訊，如 IP，但當使用 Traefik 架構後，所有 User 都是訪問 Traefik，藉由 Traefik 轉發 Request 到 API，因此這時 API 得到的 User IP，全部會變成 Traefik 的 IP，但這並不是實際上 User 的 IP，因此需要加入此設定，才能獲得 User 真實的 IP。

traefik.http.routers.api-https.rule：設定 Router 規則。Host 是前面使用 No-IP 申請的網址 testapi.ddns.net，Path 可以看成 API 的路徑。舉例來說，FastAPI 自動生成的文件，會在 /docs 路徑底下，因此需要將此規則（Host(`testapi.ddns.net`) && PathPrefix(`/docs`)），加到 Router 中，Traefik 才有辦法辨別。同理，如果讀者對於 API 有額外的版本設定，如 /api/v1、/api/v2，那 /api 這層 path，就需要加入此，Host(`testapi.ddns.net`) && PathPrefix(`/api`)。

traefik.http.services.api-https.loadbalancer.server.port=80：Traefik 的好處之一，會自動處理分流（Loading Balance），此 Port 設定為 80。

另一個不同的地方是，需要加入 traefik-public 的 Network，在同一個網路底下，才能互通，這也是 Docker 的特色之一。除了 api.yml 需要寫額外設定，因為 Traefik Loading Balance 是指定 80 Port，需要將 Dockerfile 中的 Port 改為 80，如下

https://github.com/FinMind/FinMindBook/blob/master/DataEngineering/Chapter10/10.5/api/Dockerfile

```
Dockerfile
FROM continuumio/miniconda3:4.3.27

RUN apt-get update

RUN mkdir /FinMindProject
COPY . /FinMindProject/
WORKDIR /FinMindProject/

# install package
RUN pip install pipenv && pipenv sync

# genenv
RUN VERSION=RELEASE python genenv.py

# 預設執行的指令
CMD ["pipenv", "run", "uvicorn", "api.main:app", "--host", "0.0.0.0",
"--port", "80"]
```

到這就結束了 API 與 Traefik 的結合，實際上並沒有做太多改動，但多了 SSL、自動更新 SSL 憑證、Loading Balance、Router 管控，接著開始將以上新版的 api.yml 部屬上線，指令如下

重新建立 Docker Image，linsamtw/api:10

```
docker build -f Dockerfile -t linsamtw/api:10 .
```

push

```
docker push linsamtw/api:10
```

部屬

```
docker stack deploy -c api.yml api
```

完成以上指令後，到以下網址 http://139.162.104.54:8888/dashboard/#/
http/routers

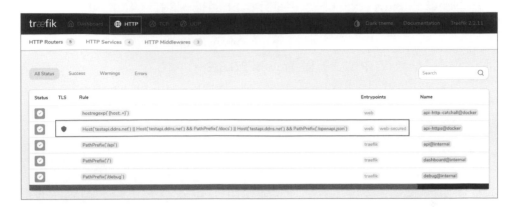

如上圖，可以觀察到，增加了新的 Rule，Host(`testapi.ddns.net`) || Host
(`testapi.ddns.net`) && PathPrefix(`/docs`) || Host(`testapi.ddns.net`) &&
PathPrefix(`/openapi.json`)，這就是前面在 api.yml 的設定，注意一點，在
TLS 有個綠色盾牌，代表成功加入 SSL 憑證，這時再回到 API 的網址，
http://testapi.ddns.net/docs

如上,從三角形驚嘆號,變成鑰鎖了,代表成功加入 SSL,而 API 的網址,也變成 **https** 了,即使輸入 http 的網址,http://testapi.ddns.net/docs,也會自動導向 https,到此已經完成了 API 網址、SSL。

讀者有興趣,可以試著將資料庫 phpMyAdmin 的頁面、Portainer 的頁面,都用網址取代 IP,試著與 Traefik 做結合。

10.6　總結

本章節使用 No-IP 申請免費網址,並使用 Traefik 管理網址、自動註冊更新 SSL 憑證,這樣對於提供 API 產品,又更進一步,實務上基本都是用網址、甚至 DNS,不太可能直接提供 IP,這對於工作或是求職上,更進一步加深技能樹。

筆者初期也是使用 Nginx 處理網址、用 Certbot 自動獲取 Let's Encrypt 憑證,但後期接觸 Docker 後,將一切都容器化,當然 Nginx 也可以用 Docker,但與專門用於集群架構的 Traefik 相比,Traefik 在管理與開發上,非常容易上手,且高拓展性,因此本書特別介紹 Traefik。讀者在 API 開發的路上,一定會遇到網址、SSL 的問題,使用 Traefik 會讓一切都變的美好。

第 4 篇
資料視覺化

11.1 什麼是視覺化？

在大數據的時代，要怎麼使用大量數據，協助商業分析、數據分析，導入數據驅動（Data Driver）的觀念，是一個蠻大的課題。以本書金融資料為例，到了第 11 章，讀者已經完成爬蟲，收集了股價資料，只要持續開發各種爬蟲，就能收集超過 10 種金融資料，收集完成後，要如何使用這些資料呢？

視覺化是一個蠻常被提起的概念，且在業界，視覺化報表是非常通用的工具，基本上有使用數據的公司，一定會導入 Dashboarsd，可能用於業務分析、流量分析、營運分析、KPI 等等，那要如何製作 Dashboard 呢？

Python 常見的繪圖工具有 matplotlib、plotly 甚至 dash，但這些都是偏向單機版，多數用於 Jupyter 上，且要寫不少程式，那有沒有更簡單的工具？可以更好的呈現給業務單位、行銷單位觀察數據呢？ Redash 就是現成的 BI（Business Intelligence）工具，下圖就是筆者使用 Redash 製作的一個範例。

上圖單純使用 Docker 架設好服務後，剩下只需要寫 SQL 即可做出 Dashboard，相較於寫 Python，開發速度快非常多，且 SQL 相對簡單，可以提供更多業務端、分析端使用，不再限於工程團隊。

可能有些讀者比較想做模型，而非視覺化，但實務上，並沒有這麼單純，不論是做模型或是資料分析，最後的結果，都需要呈現給其它部門，使用 Data 協助他們解決業務上的問題，而最好的呈現方式，就是使用 BI 工具製作 Dashboard。

且實務上並不會所有 Data 相關的問題，都需要模型，例如營運單位希望觀察每天的業績、網站流量，行銷部門需要觀察，針對不同管道投放廣告的差異，以上都會是 Data 部門負責，且都跟模型無關，這時最好的方式，就是製作 Dashboard 作呈現，以下將從架設 Redash 開始一一講解。

11.2　Redash

相信讀者看到第 11 章節後，都能猜到本書要用什麼方式去架設 Redash 了，沒錯，就是 Docker，只不過 Redash 使用的相關服務稍微多一點，根 據 Redash 官 方 GitHub，https://github.com/getredash/redash， 啟 動 Redash，至少就需要 Postgres（RDB 資料庫）、Redis（NoSQL 資料庫）、Nginx（網頁伺服器）、Worker（分散式）這四個相對應的服務，還不包含 Redash Server 本身，這也是為什麼 Redash 是業界常用的視覺化工具，因 為 Redash 將各個部分，拆解成微服務 Docker，並使用 Worker 分散式、使用 Redis 做圖表暫存，是很強大的 BI 工具。

這裡突然出現幾個名詞，讀者如果第一次看到，可能有點陌生，筆者在 此做一些簡單介紹。

Postgres，類似於 MySQL 的 RDB 資料庫，都是關聯式資料庫，主要是使 用場景不同，初學者可以把它跟 MySQL 看成一樣的東西，等遇到不同的 情境，再比較差異，基本上兩者都有不同的擁護者。

Redis，NoSQL 資料庫，最大的優勢是，資料存放在記憶體層級，使它反 應非常快速，由於網站對於反應時間有一定的要求，越快越好，因此這 裡 Redash 提供 Web 服務，一部分就是使用 Redis 存放暫存檔。

以上介紹完後，筆者開始講解如何架設 Redash，yml 檔如下：

https://github.com/FinMind/FinMindBook/blob/master/DataEngineering/ Chapter11/11.2/redash.yml

```
redash.yml
version: "3.8"
services:
```

```
redash:
  # 選用此版本的 redash image
  image: redash/redash:8.0.0.b32245
  # depends 代表依賴關係
  # 需要等 postgres、redis 先架設完成
  # 才會開始架設 redash services
  depends_on:
    - postgres
    - redis
  restart: always
  # redash 啟動 server 指令
  command: server
  # redash 使用 5000 port
  ports:
    - target: 5000
      published: 5000
      mode: host
  # 相關的環境變數，包含
  environment:
    PYTHONUNBUFFERED: 0
    # log 等級
    REDASH_LOG_LEVEL: INFO
    # redis 的連線設定
    REDASH_REDIS_URL: redis://redis:6379/0
    # postgres 的密碼
    POSTGRES_PASSWORD: postgres_password
    REDASH_COOKIE_SECRET: redash_cookie_secret
    REDASH_SECRET_KEY: redash_secret_key
    # postgres 的連線設定
    REDASH_DATABASE_URL: postgresql://postgres:postgres_password@postgres/
postgres
    # 允許在 redash 上使用 python
```

```
      REDASH_ADDITIONAL_QUERY_RUNNERS: redash.query_runner.python
      # worker 數量
      # 這是做範例，設定 1，
      # 未來可根據需求自行增加
      REDASH_WEB_WORKERS: 1
    networks:
        - my_network

  create_table:
    # 相關設定與 redash services 相同
    # 也用相同的 image
    image: redash/redash:8.0.0.b32245
    depends_on:
      - postgres
      - redis
    # restart: always
    # 主要差異在此
    # 初始化需要建立 table
    # 建立完成後，可在 portainer 上刪除此 services
    command: python /app/manage.py database create_tables
    environment:
      PYTHONUNBUFFERED: 0
      REDASH_LOG_LEVEL: INFO
      REDASH_REDIS_URL: redis://redis:6379/0
      POSTGRES_PASSWORD: postgres_password
      REDASH_COOKIE_SECRET: redash_cookie_secret
      REDASH_SECRET_KEY: redash_secret_key
      REDASH_DATABASE_URL: postgresql://postgres:postgres_password@postgres/
postgres
      REDASH_ADDITIONAL_QUERY_RUNNERS: redash.query_runner.python
      REDASH_WEB_WORKERS: 1
    networks:
```

```
      - my_network

  scheduler:
    # 用於 redash 相關的 scheduler 排程
    image: redash/redash:8.0.0.b32245
    depends_on:
      - postgres
      - redis
    restart: always
    command: scheduler
    environment:
      PYTHONUNBUFFERED: 0
      REDASH_LOG_LEVEL: INFO
      REDASH_REDIS_URL: redis://redis:6379/0
      POSTGRES_PASSWORD: postgres_password
      REDASH_COOKIE_SECRET: redash_cookie_secret
      REDASH_SECRET_KEY: redash_secret_key
      REDASH_DATABASE_URL: postgresql://postgres:postgres_password@postgres/
postgres
      REDASH_ADDITIONAL_QUERY_RUNNERS: redash.query_runner.python
      # 這裡的 celery 就跟前面章節提到的
      # 分散式概念一樣
      QUEUES: "celery"
      WORKERS_COUNT: 1
    networks:
      - my_network

  scheduled_worker:
    # 分散式中的 worker
    image: redash/redash:8.0.0.b32245
    depends_on:
      - postgres
```

```
    - redis
  restart: always
  command: worker
  environment:
    PYTHONUNBUFFERED: 0
    REDASH_LOG_LEVEL: INFO
    REDASH_REDIS_URL: redis://redis:6379/0
    POSTGRES_PASSWORD: postgres_password
    REDASH_COOKIE_SECRET: redash_cookie_secret
    REDASH_SECRET_KEY: redash_secret_key
    REDASH_DATABASE_URL: postgresql://postgres:postgres_password@postgres/
postgres
    REDASH_ADDITIONAL_QUERY_RUNNERS: redash.query_runner.python
    # 負責處理 queue = scheduled_queries,schemas 類型的任務
    # 與本書中 7.6.3，crawler.yml 中的 -Q twse 概念一樣
    QUEUES: "scheduled_queries,schemas"
    WORKERS_COUNT: 1
  networks:
    - my_network

adhoc_worker:
  image: redash/redash:8.0.0.b32245
  depends_on:
    - postgres
    - redis
  restart: always
  command: worker
  environment:
    PYTHONUNBUFFERED: 0
    REDASH_LOG_LEVEL: INFO
    REDASH_REDIS_URL: redis://redis:6379/0
    POSTGRES_PASSWORD: postgres_password
```

```
    REDASH_COOKIE_SECRET: redash_cookie_secret
    REDASH_SECRET_KEY: redash_secret_key
    REDASH_DATABASE_URL: postgresql://postgres:postgres_password@postgres/
postgres
    REDASH_ADDITIONAL_QUERY_RUNNERS: redash.query_runner.python
    # 負責處理 queue = queries 類型的任務
    # 與本書中 7.6.3，crawler.yml 中的 -Q twse 概念一樣
    QUEUES: "queries"
    WORKERS_COUNT: 2
  networks:
    - my_network

 redis:
  # redis 資料庫
  image: redis:5.0
  restart: always
  volumes:
    - redis-redash:/bitnami/redis/data
  networks:
    - my_network

 postgres:
  # postgres 資料庫
  image: postgres:9.6-alpine
  environment:
    PYTHONUNBUFFERED: 0
    REDASH_LOG_LEVEL: INFO
    REDASH_REDIS_URL: redis://redis:6379/0
    POSTGRES_PASSWORD: postgres_password
    REDASH_COOKIE_SECRET: redash_cookie_secret
    REDASH_SECRET_KEY: redash_secret_key
    REDASH_DATABASE_URL: postgresql://postgres:postgres_password@postgres/
```

```
postgres
    REDASH_ADDITIONAL_QUERY_RUNNERS: redash.query_runner.python
  volumes:
    - postgres:/var/lib/postgresql/data
  restart: always
  networks:
      - my_network

nginx:
  # 網頁伺服器
  image: redash/nginx:latest
  depends_on:
    - redash
  restart: always
  networks:
      - my_network

networks:
  my_network:
    external: true

volumes:
  postgres:

  redis-redash:
```

在部屬之前，有一點需要注意，一開始本書使用的雲端 Linode，是最低
方案，只有 1GB ram，由於 Redash 需要多個服務，因此麻煩讀者將機器
升級到 4GB ram 方案，20 美金／月（如果是使用個人電腦的讀者，則無
此問題）。

部屬指令如下：

```
docker stack deploy -c redash.yml redash
```

成功執行後，畫面如下，建立了 7 個 Service：

```
sam@localhost:~/FinMindBook/DataEngineering/Chapter11/11.2$ docker stack deploy -c redash.yml redash
Ignoring unsupported options: restart

Creating service redash_scheduler
Creating service redash_scheduled_worker
Creating service redash_adhoc_worker
Creating service redash_redis
Creating service redash_postgres
Creating service redash_nginx
Creating service redash_redash
Creating service redash_create_table
sam@localhost:~/FinMindBook/DataEngineering/Chapter11/11.2$
```

接著轉到 Portainer 的頁面，如下圖，在 1 的地方，輸入 redash 做 filter，
之後在 2 的地方，可以看到前面建立的 redash 服務，Port 是 5000：

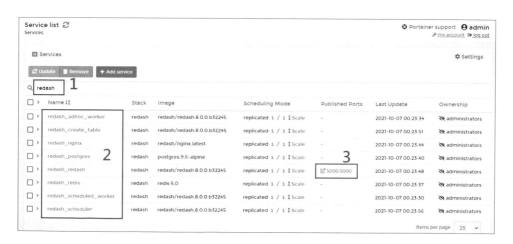

接著到以下網址 http://139.162.104.54:5000（IP 請自行換成讀者的 Linode
網址），如果看到以下畫面，代表部屬成功：

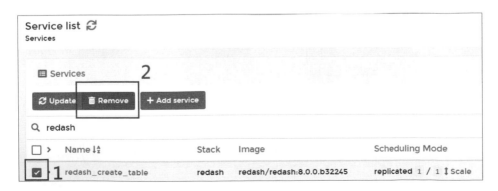

這時就可以把 create_table 的 service 移除了，方法如下：

這時就完成了 Redash 的部屬了，是不是很簡單，讀者可搭配 Chapter 10 的 Traefik，幫你的 Redash 申請網址，下一章節，將講解如何使用 Redash。

11.3　Redash 帳號設定

首先，先進行帳號設定，如下圖

Redash Initial Setup

Welcome to Redash!
Before you can use your instance, you need to do a quick setup.

Admin User　**1**
Name
　sam

Email Address
　samlin266118@gmail.com

Password
　•••••••••

☑ Subscribe to Security Notifications
☑ Subscribe to newsletter (version updates, no more than once a month)

General
Organization Name　**2**
　FinMind

Used in email notifications and the UI.

　3
Setup

在 1 的地方，輸入你的名稱、Email、密碼，並在 2 的地方，輸入你的組織名稱，隨便你喜歡輸入什麼都可以，最後在 3 點選 Setup。接著就會進入以下畫面：

如果想幫團隊成員，開設帳號的話，可以按照以下步驟，點選右上角自己的圖示，選取 Users，最後到 3 的地方，New User，就可以新增使用者啦，非常的方便。

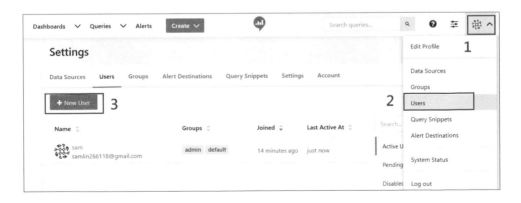

下一章，介紹如何與 MySQL 資料庫做連接，拿取金融資料做視覺化。

11.4　資料庫連接

如下圖，點選右上角圖示，再選取 Data Sources：

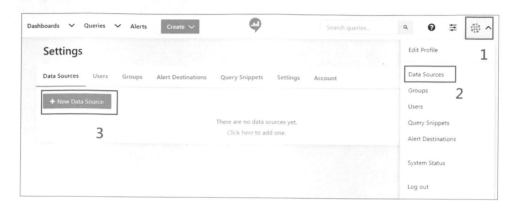

到以下畫面後，可以看到有超過 20 種的 DataBase 可以選擇，包含知名的 BigQuery，有興趣讀者可再自行研究，這裡筆者使用 MySQL，在 1 的地方輸入 MySQL 後，點選 MySQL：

會再看到以下畫面，就一一輸入各個設定，Name、Host、Port、User、Password、Database name，特別注意一點，Host 的地方，一般來說是輸入 IP，但因為本書中在 Swarm 架構下，且跟 MySQL 統一個 Network，因此 Host 輸入 Service Name - mysql，是可以的

11.5 匯入資料

在製作視覺化儀表板之前，麻煩讀者先自行收集，台股股價、三大法人買賣、融資融券、外資持股等資料，有資料才有辦法做視覺化。收集方法也很簡單，只要將章節 8.1.4 的部分，照著模仿股價爬蟲，因為證交所爬蟲架構都非常類似，可以輕易地複製出三大法人、融資融券、外資持股爬蟲等。

由於資料收集會花費許多時間，本書先將所需資料，上傳雲端，並提供程式碼，讓讀者可以直接將資料上傳到讀者架設的資料庫中，如下。

https://github.com/FinMind/FinMindBook/blob/master/DataEngineering/Chapter11/11.5/upload_data2mysql.py

```
upload_data2mysql.py
import os
import sys


import pandas as pd
from loguru import logger
from sqlalchemy import (
    create_engine,
    engine,
)
from tqdm import tqdm


import wget



def get_mysql_financialdata_conn() -> engine.base.Connection:
    # TODO 請將 IP 換成讀者自己的 IP
```

```python
    address = "mysql+pymysql://root:test@139.162.104.54:3306/FinancialData"
    engine = create_engine(address)
    connect = engine.connect()
    return connect



def create_taiwan_stock_info_sql():
    return """
        CREATE TABLE `taiwan_stock_info` (
            `industry_category` varchar(32) CHARACTER SET utf8 COLLATE
utf8_unicode_ci NOT NULL,
            `stock_id` varchar(32) CHARACTER SET utf8 COLLATE utf8_
unicode_ci NOT NULL,
            `stock_name` varchar(30) CHARACTER SET utf8 COLLATE utf8_
unicode_ci DEFAULT NULL,
            `type` varchar(4) CHARACTER SET utf8 COLLATE utf8_unicode_ci
NOT NULL COMMENT '上市twse/上櫃tpex',
            `date` date DEFAULT NULL,
            PRIMARY KEY (`stock_id`,`industry_category`)
        ) ENGINE=InnoDB DEFAULT CHARSET=utf8mb3 COLLATE=utf8_unicode_ci;
    """



def create_taiwan_stock_price_sql():
    return """
        CREATE TABLE `taiwan_stock_price` (
            `StockID` varchar(10) COLLATE utf8_unicode_ci NOT NULL,
            `Transaction` bigint NOT NULL,
            `TradeVolume` int NOT NULL,
            `TradeValue` bigint NOT NULL,
            `Open` float NOT NULL,
            `Max` float NOT NULL,
            `Min` float NOT NULL,
```

```
            `Close` float NOT NULL,
            `Change` float NOT NULL,
            `Date` date NOT NULL,
            PRIMARY KEY(`StockID`, `Date`)
        ) ENGINE=InnoDB DEFAULT CHARSET=utf8mb3 COLLATE=utf8_unicode_ci
        PARTITION BY KEY (StockID)
        PARTITIONS 10;
    """


def create_taiwan_stock_institutional_investors_sql():
    return """
        CREATE TABLE taiwan_stock_institutional_investors(
            `name` VARCHAR(20),
            `buy` BIGINT(64),
            `sell` BIGINT(64),
            `stock_id` VARCHAR(10),
            `date` DATE,
            PRIMARY KEY(`stock_id`, `date`, `name`)
        ) PARTITION BY KEY(`stock_id`) PARTITIONS 10;
    """


def create_taiwan_stock_margin_purchase_short_sale_sql():
    return """
        CREATE TABLE `taiwan_stock_margin_purchase_short_sale`(
            `stock_id` VARCHAR(10) CHARACTER SET utf8 COLLATE utf8_
unicode_ci NOT NULL COMMENT '股票代碼',
            `MarginPurchaseBuy` BIGINT NOT NULL COMMENT '融資買進',
            `MarginPurchaseSell` BIGINT NOT NULL COMMENT '融資賣出',
            `MarginPurchaseCashRepayment` BIGINT NOT NULL COMMENT
'融資現金償還',
```

```
            `MarginPurchaseYesterdayBalance` BIGINT NOT NULL COMMENT
'融資昨日餘額',
            `MarginPurchaseTodayBalance` BIGINT NOT NULL COMMENT
'融資今日餘額',
            `MarginPurchaseLimit` BIGINT NOT NULL COMMENT '融資限額',
            `ShortSaleBuy` BIGINT NOT NULL COMMENT '融券買進',
            `ShortSaleSell` BIGINT NOT NULL COMMENT '融券賣出',
            `ShortSaleCashRepayment` BIGINT NOT NULL COMMENT '融券償還',
            `ShortSaleYesterdayBalance` BIGINT NOT NULL COMMENT '融券昨日
餘額',
            `ShortSaleTodayBalance` BIGINT NOT NULL COMMENT '融券今日餘額',
            `ShortSaleLimit` BIGINT NOT NULL COMMENT '融券限制',
            `OffsetLoanAndShort` BIGINT DEFAULT NULL COMMENT '資券互抵',
            `date` DATE NOT NULL COMMENT '日期',
            PRIMARY KEY(`stock_id`, `date`)
    ) PARTITION BY KEY(`stock_id`) PARTITIONS 10;
    """


def create_taiwan_stock_holding_shares_per_sql():
    return """
        CREATE TABLE taiwan_stock_holding_shares_per (
            `HoldingSharesLevel` VARCHAR(19),
            `people` INT(10),
            `unit` BIGINT(64),
            `percent` FLOAT,
            `stock_id` VARCHAR(10),
            `date` DATE,
            `update_time` DATETIME NOT NULL DEFAULT CURRENT_TIMESTAMP,
            PRIMARY KEY (`stock_id`,`date`,`HoldingSharesLevel`)
        )
        PARTITION BY KEY(stock_id)
```

```
        PARTITIONS 10;
    """


def create_table(table: str):
    mysql_conn = (
        get_mysql_financialdata_conn()
    )
    sql = eval(f"create_{table}_sql()")
    try:
        logger.info(
            f"create table {table}"
        )
        mysql_conn.execute(sql)
    except:
        logger.info(
            f"{table} already exists"
        )


def download_data(table: str):
    logger.info("download data")
    if f"{table}.csv" in os.listdir(
        "."
    ):
        logger.info(f"already download")
    else:
        url = f"https://github.com/FinMind/FinMindBook/releases/download/
data/{table}.csv"
        wget.download(
            url, f"{table}.csv"
        )
```

```
        logger.info(
            "download data complete"
        )


def upload_data2mysql(table: str):
    chunk_size = 100000
    mysql_conn = (
        get_mysql_financialdata_conn()
    )
    try:
        logger.info("load data")
        logger.info("upload to mysql")
        reader = pd.read_csv(
            f"{table}.csv",
            chunksize=chunk_size,
        )
        for df_chunk in tqdm(reader):
            df_chunk.to_sql(
                name=table,
                con=mysql_conn,
                if_exists="append",
                index=False,
            )
    except:
        logger.info("already upload")


def main(table: str):
    create_table(
        table=table,
    )
```

```
    download_data(
        table=table,
    )
    upload_data2mysql(table=table)



if __name__ == "__main__":
    table = sys.argv[1]
    main(table)
```

其中需注意一點，把 MySQL 的 IP，139.162.104.54，換成讀者自己的，才能成功上傳資料。

以上包含台股基本資料、台股股價、三大法人、融資融券資料，使用方式如下

下載台股基本資料

```
pipenv run python upload_data2mysql.py taiwan_stock_info
```

下載台股股價

```
pipenv run python upload_data2mysql.py taiwan_stock_price
```

下載台股三大法人

```
pipenv run python upload_data2mysql.py taiwan_stock_institutional_
investors
```

下載台股融資融券

```
pipenv run python upload_data2mysql.py taiwan_stock_margin_purchase_
short_sale
```

下載台股股東持股分級表

```
pipenv run python upload_data2mysql.py taiwan_stock_holding_shares_per
```

以上指令會下載資料，這需要一段時間，完成後，資料庫會如下圖：

這時資料庫都有資料後，下一章節，開始製作第一個圖表。

11.6 製作第一個圖表

首先，到 Redash 的頁面，根據下圖操作，建立新的 Query：

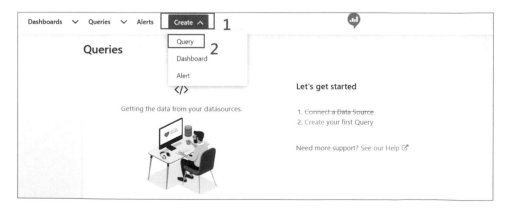

要如何建立新的 Query 呢?請讀者在以下圖表 1,輸入以下 SQL,撈出 2330 的收盤價與日期:

https://github.com/FinMind/FinMindBook/blob/master/DataEngineering/ Chapter11/11.6/stock_price.sql

```
SELECT Close, Date
FROM taiwan_stock_price
WHERE StockID = '2330'
order by Date
```

完成輸入 SQL 後,請讀者再並按下 Save 儲存,與 Execute 執行:

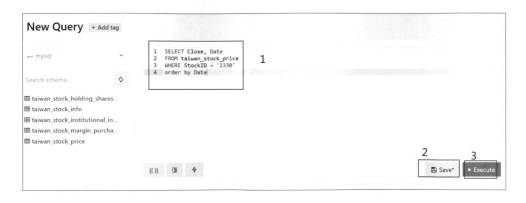

Redash 會根據以上 SQL,去 MySQL 下指令,撈對應的資料回來,結果如下:

一開始只會回傳 table，如果想要做視覺化，請點選 4，New Visualization

會看到上圖，Chart Type 代表圖形種類，因為是股價，選擇 Line，畫折線圖，而一般折線圖，X 軸會是 Date，在 X Column 選擇 Date，Y 軸是股價，在 Y Columns 選擇 Close：

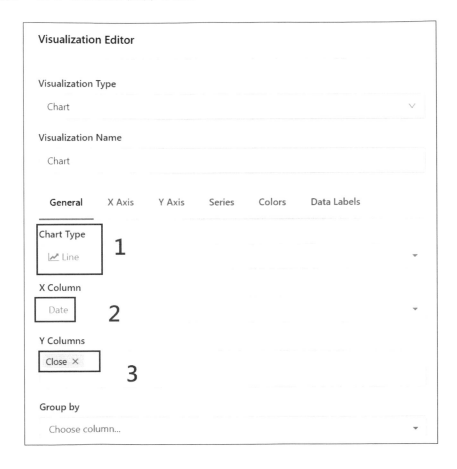

以下就是 Redash 畫出的折線圖，再點選 Save 儲存這張視覺化圖表：

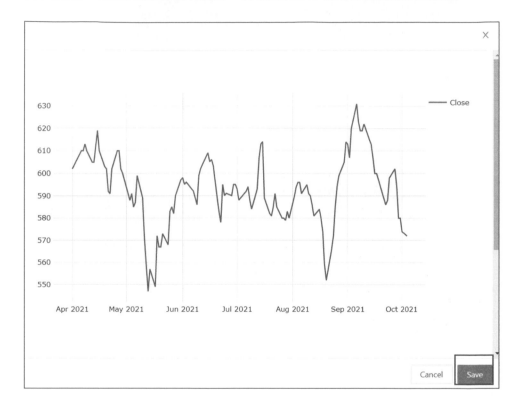

是不是很簡單？當讀者收集完資料後，後續只需要寫簡單的 SQL，就能做出視覺化圖表。但只看股價太單純了，進階一點的應用，想知道其他資料與股價的相關性，下一個圖表，本書串接股價與外資買賣的圖，觀察外資是否與股價有相關性。

步驟與創建股價 Query 相同，只需更改 SQL，SQL 如下：

https://github.com/FinMind/FinMindBook/blob/master/DataEngineering/
Chapter11/11.6/stock_price_vs_foreign_investor.sql

```
SELECT if(tsii.buy-tsii.sell>0, tsii.buy-tsii.sell,0) AS '買超',
       if(tsii.sell-tsii.buy>0, tsii.sell-tsii.buy,0) AS '賣超',
       stock_price.Close AS '股價',
       tsii.date AS date
FROM taiwan_stock_institutional_investors AS tsii
INNER JOIN taiwan_stock_info AS si ON si.stock_id = tsii.stock_id
INNER JOIN taiwan_stock_price AS stock_price ON tsii.stock_id =
stock_price.StockID
AND tsii.date = stock_price.Date
WHERE tsii.name = 'Foreign_Investor'
  AND si.stock_name = '台積電'
  AND tsii.date >= '2015-01-01'
```

以上 SQL 將股價與三大法人做 **Join**，才能同時顯示，**外資買賣**與**股價**，
且同時 Join 台股資訊，因為相較於股票代碼，股票名稱比較好讀。這裡
不單純看外資買賣的量，而是進而看買賣超的數據，因此對於買賣資料
也進行額外的計算處理。

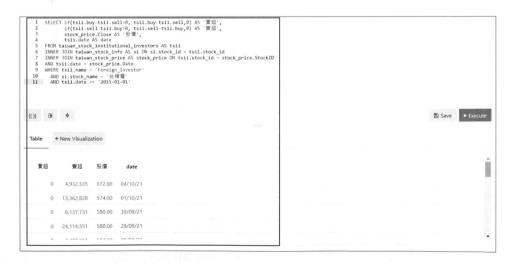

成功從 MySQL 撈資料後，接著點選 New Visualization，做圖：

Visualization Editor

Visualization Type

| Chart | ⌄ |

Visualization Name

| Chart |

General   X Axis   Y Axis   Series   Colors   Data Labels

Chart Type

| 📈 Line | ▾ |

X Column

| date | ▾ |

Y Columns

| 買超 ✕   賣超 ✕   股價 ✕ |

Group by

| Choose column... | ▾ |

Errors column

| Choose column | ▾ |

Chart Type 同樣選 Line，X Column 選 date，Y Columns 選擇買超、賣超、股價，呈現的圖表如下：

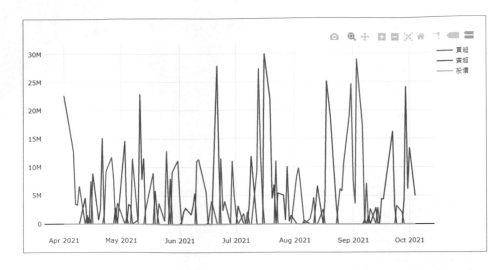

因為買賣超資料量尺與股價不同，台積電股價約在 700 以下，但買賣超動輒百萬，需要做後續處理。點選 Series，將不同數值選擇要使用 Y 左軸或是右軸，本書將買賣超選擇 Left Y Axis，股價選擇 Right Y Axis，並將買賣超的 Type，改成 Bar

調整完後的圖表如下，相對來說，比較好閱讀。

接著再調整顏色，一般而言，紅色代表買超、綠色代表賣超，設定如下，在 Colors 的部分去設定顏色，買超設定紅色，賣超設定綠色，股價就選擇藍色。

修改完後的圖表如下，是不是更好理解了，在股價大幅下降的期間，外資大量賣超，相反，股價大幅上漲，剛好就是外資大量買超，最後點選右下角的 Save，將此圖存檔。

最後到左上角的位置，設定這個 Query 的名稱，本書設定是，股價 vs 外資：

讀者可根據以上方法，額外增加，投信 vs 股價，SQL 如下，就不再帶著
讀者一一設定了，

```
SELECT if(tsii.buy-tsii.sell>0, tsii.buy-tsii.sell,0) AS '買超',
       if(tsii.sell-tsii.buy>0, tsii.sell-tsii.buy,0) AS '賣超',
       stock_price.Close AS '股價',
       tsii.date AS date
FROM taiwan_stock_institutional_investors AS tsii
INNER JOIN taiwan_stock_info AS si ON si.stock_id = tsii.stock_id
INNER JOIN taiwan_stock_price AS stock_price ON tsii.stock_id = stock_
price.StockID
AND tsii.date = stock_price.Date
WHERE tsii.name = 'Investment_Trust'
  AND si.stock_name = '台積電'
  AND tsii.date >= '2015-01-02'
```

股價 vs 投信如下：

以上完成兩張圖表後，要如何在同一的畫面，看到多張圖表呢？下一章
節，本書將介紹 dashboard，可以組合多種圖表。

11.7　第一個 Dashboard

前一章節，製作多個 Query，多種資料的視覺化，而 Dashboard，就是將多種 Query 的視覺化圖表，組合起來，方法如下，點選 Create，這次選擇 Dashboard。

名稱部分，讀者可自行設定，本章節以籌碼視覺化為主，因此這邊使用籌碼分析。

設定好名稱後，會進入以下畫面，點選右下角的 Add Widget，將前一章節的圖表加進來。

可以看到，以下三個，股價 vs 投信、股價 vs 外資、股價，剛好對應到前一章節設定的三個 Query，這裡選擇股價 vs 投信。

Choose Visualization 的 意 思 是，選 擇 視 覺 化 種 類，分 別 有 Table 與
Chart，也對應到在 Query 開發中的 Table 與 Chart，本書在此選擇 Chart
圖表，之後點選 Add to Dashboard，加到 Dashboard 上。

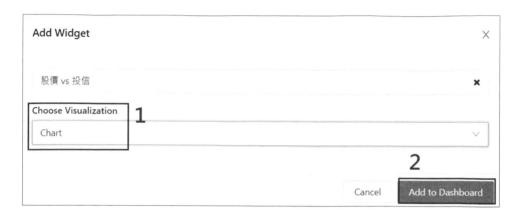

結果如下圖，成功將前一章節設定的 Query 圖表，加進 Dashboard 啦。

同樣的操作，將股價 vs 外資也加進來：

最後點選右上角，Done Editing，完成編輯，到此就完成第一個 Dashboard 了。

但目前為止，只能觀看台積電的分析，要如何輕易觀看其他股票呢？是否有一個地方，可以讓讀者直接選取想看得股票呢？以下筆者一一介紹

11.8 設定下拉式選單

第一步，先建立股票清單，使用 Query，SQL 如下，因為筆者只操作非權證股票，因此以下 SQL 將權證類型的股票排除掉：

https://github.com/FinMind/FinMindBook/blob/master/DataEngineering/
Chapter11/11.6/stock_name_list.sql

```
SELECT DISTINCT
    `stock_name`
FROM
```

```
    `taiwan_stock_info`
WHERE
    (
        `industry_category` NOT IN(
            '認購權證(不含牛證)',
            '可展延牛證',
            '認售權證(不含熊證)',
            '熊證(不含可展延熊證)',
            '牛證(不含可展延牛證)',
            '不動產投資信託證券',
            '上櫃指數股票型基金(ETF)',
            '受益證券',
            '指數投資證券(ETN)',
            'ETN',
            '存託憑證',
            '封閉式基金'
        )
    )
```

將左上角的 Query 名稱，設為 stock name list，並點選右上角圖 3 的
Publish，將此 Query 設為公開，如果不設公開，後續無法使用。

需要注意一點，筆者在此建立的是**股票名稱**清單，而非**股票代碼**，因為一般人相較於股票代碼，股票名稱比較好理解。

完成後，到股價 vs 外資的 Query 頁面，在下圖 1 的地方，將原本台積電的位置，改成，{{ 股票名稱 }}，這時 Redash 在圖 2 的地方，突然出現相對應的股票名稱，這有什麼用途呢？

新長出來的部分，本書將會用此來設定下拉式選單，請讀者先點選 3 的地方，一個齒輪。

點選後，會跳出一個視窗，接著在點上圖 1，Text，會出現好幾個選項，分別代表什麼意思呢？

Text：可以輸入任意文字。

Number：顧名思義，輸入數字。

Dropdown List：提供下拉式選單列表，而選項就是讀者輸入的列表，例如讀者輸入，a,b,c，這三個選項，如下：

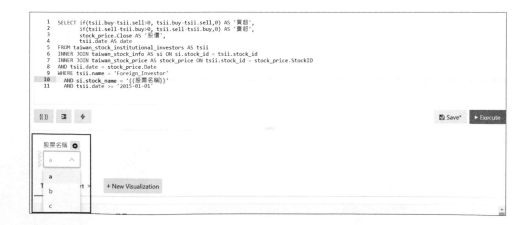

之後回到原先 Query 的畫面，就變成下拉式選單了。

```
1  SELECT if(tsii.buy-tsii.sell>0, tsii.buy-tsii.sell,0) AS '買超',
2      if(tsii.sell-tsii.buy>0, tsii.sell-tsii.buy,0) AS '賣超',
3      stock_price.Close AS '股價',
4      tsii.date AS date
5  FROM taiwan_stock_institutional_investors AS tsii
6  INNER JOIN taiwan_stock_info AS si ON si.stock_id = tsii.stock_id
7  INNER JOIN taiwan_stock_price AS stock_price ON tsii.stock_id = stock_price.StockID
8  AND tsii.date = stock_price.Date
9  WHERE tsii.name = 'Foreign_Investor'
10     AND si.stock_name = '{{股票名稱}}'
11     AND tsii.date >= '2015-01-01'
```

Query Based Dropdown List：顧名思義，根據 Query 的結果，產生的下拉
式選單，與前一個 Dropdown List 相比，不需要一個個輸入選項，Redash
會直接讀取 Query 的結果，作為選項。

Date：日期欄位。

Date and Time：日期與時間欄位。

Date Range：日期區間欄位。

Date and Time Range：日期與時間區間欄位。

此章節需要下拉式選單，因此選擇 Query Based Dropdown List，之後在
Query 的地方，選擇 stock name list，最後點選 OK。

設定完成後，這時股票名稱，就成功變成下拉式選單了。

設定完股票名稱的下拉式選單後，如果想看不同時間區間，怎麼設定呢？

需要使用 **Date Range** 選項，方法如下，將日期的地方，改成，{{date. start}} 與 {{date.end}}，Date Range 的特性是，第一個日期是 date.start，第二個日期是，date.end，使用此方法，可以拉取任意時間區間：

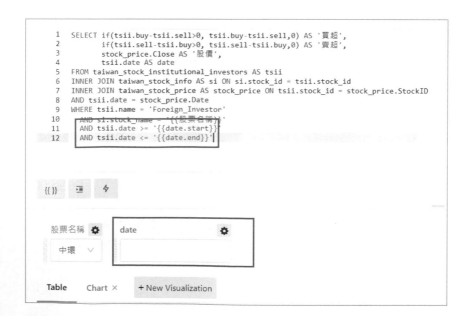

同樣的方式，點選 date 右上角的齒輪後，在 Type 的地方選取 Date Range

這時就可以開始選取時間區間了。

本章節到此，已經將股價 vs 外資部分的 Query，完成下拉式選單，與時間區間的設定，麻煩讀者照著相同步驟，將股價 vs 投信的 Query 也做同樣的設定，可參考以下 SQL：

https://github.com/FinMind/FinMindBook/blob/master/DataEngineering/
Chapter11/11.6/stock_price_vs_investment_trust_with_params.sql

```
SELECT if(tsii.buy-tsii.sell>0, tsii.buy-tsii.sell,0) AS '買超',
       if(tsii.sell-tsii.buy>0, tsii.sell-tsii.buy,0) AS '賣超',
       stock_price.Close AS '股價',
       tsii.date AS date
FROM taiwan_stock_institutional_investors AS tsii
INNER JOIN taiwan_stock_info AS si ON si.stock_id = tsii.stock_id
INNER JOIN taiwan_stock_price AS stock_price ON tsii.stock_id =
stock_price.StockID
AND tsii.date = stock_price.Date
WHERE tsii.name = 'Investment_Trust'
  AND si.stock_name = '{{股票名稱}}'
  AND tsii.date >= '{{date.start}}'
  AND tsii.date <= '{{date.end}}'
```

完成 Query 的設定後，開始設定 Dashboard 上的下拉式選單、日期選單，方法如下：

先跳轉到 Dashboard，籌碼分析的頁面。

可以看到，兩張 Chart 都有相對應的下拉式選單、日期選單，但一個一個選取太笨了，有沒有辦法，設定 Dashboard 的參數呢？

先到股票 vs 投信的地方，點選右上角，三個點的地方後，選取 Edit
Parameters。

點選下圖的鉛筆。

將股票名稱與 date 的 Widget parameter，都改為 New dashboard parameter

之後下拉式選單與日期選單，就往上移，變成 dashboard 的參數了，這裡類似全域變數的概念，同樣的步驟，將股價 vs 外資的參數也做修改，這時要特別注意一點，前面使用的選項是 New dashboard parameter，新增一個 Dashboard 的參數，但在股價 vs 外資，就不能選擇 New dashboard parameter，因為此參數已經存在，因此選擇第二個，Existing dashboard parameter，使用現存的 Dashboard 參數。

最後，兩個 Chart 的參數，都變成 Dashboard 的參數了。

完成此步驟後，就可以任意對不同股票，切換圖表進行籌碼分析。

籌碼分析 Dashboard，目前只使用外資、投信，而在 11.5 章節中，本書下載了三大法人、融資融券、持股分級表，請讀者接續著將其他資料，都做到 Dashboard 上，整張 Dashboard 就會變得更加豐富，有助於後續分析。甚至照著本書方法，增加其他爬蟲，收集其他資料，將技術分析、籌碼分析、基本面分析，都做成不同的 Dashboard，製作專屬的看盤工具。

11.9　其他 BI 工具

市面上還有其他類似的 BI 工具，都與 Redash 類似，學會一套後，即使工作上使用其他 BI 工具，也能輕易上手，概念都是使用 SQL 撈資料，操作 UI 設計圖表，多數使用數據做分析的公司，基本上都會使用 BI 工具做分析，因此該項技能，對讀者的職涯很有幫助，以下介紹幾個常見的 BI 工具。

Superset：https://github.com/apache/superset，由 Apache 基金會開發的視覺化工具，在 Github 上有超過 40K stars，相較於 Redash 約 20K stars，這更多人使用。與 Redash 相比，有更好的權限控管機制。在公司中，不是每個人都能看到所有的 Dashboard，例如營收相關資訊，就會限定只有特定人士才能觀看，這時 BI 工具，就需要更細部的權限控管工具。因此，不同 BI 視覺化工具，根據讀者的需求選用即可。

Grafana：https://github.com/grafana/grafana，在 GitHub 上有超過 40 K stars，另一套視覺化工具，主要用於監控系統，製作監控儀表板，同樣可串接各種 ChatBot。

本書以**真實案例**出發,一步步介紹所使用的工具,與對應的場景,不只是單純介紹工具,真實案例,有助於讀者理解,這些工具是為了解決什麼問題,之後面對其他真實問題,可以很輕易地做出架構圖,每個問題,讀者都有相對應的工具。

以本書來說:

1. 筆者初期做大數據分析,需要大量數據,為了解決 data 問題,開始開發爬蟲,自動化收集資料。

2. 在爬蟲過程中,遇到被 ban IP 問題,開始接觸分散式工具 RabbitMQ、Flower、Celery,分散到多台機器上,多個 IP,成功解決被 ban IP 問題。

3. 分散式資料收集，每台機器各自收集資料，資料當然不可能存成 csv 等檔案，為了解決資料存放問題，開始架設 MySQL 資料庫，統一將爬蟲資料，存放到資料庫。

4. 分散式爬蟲使用了 RabbitMQ、Flower 、MySQL 等多個服務，且分散式要部屬到多台機器上，過程複雜，因此使用 Docker 解決架設服務問題，且解決多機器部屬更新問題。

5. 使用 Docker 後，為了解決管理多個 Docker 問題，開始接觸 Docker Swarm，並使用 Portainer 管理介面，統一管理多台機器。

6. 資料收集完成後，要如何給其他人使用？這時開始接觸 API，讓任何語言都能輕易串接。

7. 有 API 後，單純給對方 IP 讓對方發送 request 嗎？太陽春，經過前面爬蟲的觀察後，多數網站的 API 都是使用網址，為了解決網址問題，開始接觸 No-IP，申請免費 IP。

8. API 有網址後，卻缺少 SSL，導致網站安全性低，開始接觸免費 SSL 憑證 Let's Encrypt。

9. 要讓 API 使用 SSL 憑證，多數使用 Nginx，但現代架構，跟容器技術結合，因此選用 Traefik，功能更齊全。

10. 資料收集、API，都只停留在資料工程面，要如何應用呢？開始架設視覺化工具 Redash，並製作 Dashboard，讓資料產品，跨出第一步，有視覺化分析工具，非常有助於資料分析產品化，不再只停留在工程面。

11. 其中沒提到的部分，CICD，是為了解決，多人團隊開發上的問題，在每次的 PR 都做測試，並自動化部屬過程，讓團隊開發上，更加順暢、穩健。

以上步驟，從資料收集開始，到分散式爬蟲、資料庫、API、Docker、CICD、Redash，成功從 0 到 1 打造出一套資料分析產品。讀者可以根據自己有興趣的領域，做爬蟲收集資料，最後做出視覺化報表，完成專屬個人的分析作品，體驗產品開發的過程。

例如對各大求職網站上，對職缺、薪資進行分析。
對各大電商平台，產品價格、產品種類分析。
對各大社群平台，分析群眾。

實際上，有非常多題目可以做，不用愁要選什麼題目做 Side Project，當你擁有以上技能後，要做什麼都不是問題。也可以跟筆者一樣，將資料做成 API、Open Data 方式，放到 GitHub 上，提供大眾使用。

如果對於本書有疑問，可以寄信到以下信箱，samlin266118@gmail.com，或是到 FB 搜尋 FinMind 社團，https://www.facebook.com/groups/401634838071226，筆者會一一回覆你。

Note

Note